AF780139

Applications of Ferrites

Edited by Maaz Khan

Published in London, United Kingdom

Applications of Ferrites
http://dx.doi.org/10.5772/intechopen.1000321
Edited by Maaz Khan

Contributors
Sumayya M. Ansari, Vikas Kashid, Bhavesh B. Sinha, Debasis Sen, Yesh D. Kolekar, Chintalapalle V. Ramana, Amina Ibrahim Ghoneim, Tonmoye Sarkar Shathi, Abdur Rahman, Jnanranjan Panda, Bhabani Sankar Satapathy, Abhishek Mishra, Biswabhusan Biswal, Pralaya Kumar Sahoo, Herlinda Montiel, José Francisco Barrón-López, Guillermo Alvarez, Md. Ziaul Ahsan

© The Editor(s) and the Author(s) 2024

The rights of the editor(s) and the author(s) have been asserted in accordance with the Copyright, Designs and Patents Act 1988. All rights to the book as a whole are reserved by INTECHOPEN LIMITED. The book as a whole (compilation) cannot be reproduced, distributed or used for commercial or non-commercial purposes without INTECHOPEN LIMITED's written permission. Enquiries concerning the use of the book should be directed to INTECHOPEN LIMITED rights and permissions department (permissions@intechopen.com).

Violations are liable to prosecution under the governing Copyright Law.

Individual chapters of this publication are distributed under the terms of the Creative Commons Attribution 3.0 Unported License which permits commercial use, distribution and reproduction of the individual chapters, provided the original author(s) and source publication are appropriately acknowledged. If so indicated, certain images may not be included under the Creative Commons license. In such cases users will need to obtain permission from the license holder to reproduce the material. More details and guidelines concerning content reuse and adaptation can be found at http://www.intechopen.com/copyright-policy.html.

Notice

Statements and opinions expressed in the chapters are these of the individual contributors and not necessarily those of the editors or publisher. No responsibility is accepted for the accuracy of information contained in the published chapters. The publisher assumes no responsibility for any damage or injury to persons or property arising out of the use of any materials, instructions, methods or ideas contained in the book.

First published in London, United Kingdom, 2024 by IntechOpen
IntechOpen is the global imprint of INTECHOPEN LIMITED, registered in England and Wales, registration number: 11086078, 5 Princes Gate Court, London, SW7 2QJ, United Kingdom

British Library Cataloguing-in-Publication Data
A catalogue record for this book is available from the British Library

Additional hard and PDF copies can be obtained from orders@intechopen.com

Applications of Ferrites
Edited by Maaz Khan
p. cm.
Print ISBN 978-1-83769-944-5
Online ISBN 978-1-83769-943-8
eBook (PDF) ISBN 978-1-83769-945-2

We are IntechOpen,
the world's leading publisher of
Open Access books
Built by scientists, for scientists

6,800+
Open access books available

183,000+
International authors and editors

200M+
Downloads

156
Countries delivered to

Our authors are among the
Top 1%
most cited scientists

12.2%
Contributors from top 500 universities

WEB OF SCIENCE™

Selection of our books indexed in the Book Citation Index
in Web of Science™ Core Collection (BKCI)

Interested in publishing with us?
Contact book.department@intechopen.com

Numbers displayed above are based on latest data collected.
For more information visit www.intechopen.com

Meet the editor

Dr. Maaz Khan is a senior researcher at the Pakistan Institute of Nuclear Science and Technology. He completed his Ph.D. and postdoctoral research in the field of material science. His research interests include the fabrication of nanomaterials and their structural, optical, magnetic, and electrical characterizations. He has authored more than 140 research articles and published 12 books. He is currently editor-in-chief of *The Nucleus* and *Journal of Materials, Processing, and Design*, and executive editor of the *International Journal of Nano Studies and Technology*. He also serves as an editorial board member for several other materials science journals.

Contents

Preface

Ferrites are ceramic materials composed of iron oxide mixed with small quantities of one or more metallic elements, such as cobalt, nickel, zinc, manganese, strontium, and barium. There are different types of ferrites, including spinel, garnet, ortho, and hexagonal ferrites. The crystallography, structural, electrical, and magnetic properties of ferrites depend upon the fabrication, and chemical composition, as well as the various heat treatments during the preparation of ferrites. Ferrites have unique electrical and magnetic characteristics that are useful for a wide range of technological applications, including refrigerators, air conditioners, microwave ovens, radio and telecommunication devices, and computers.

This book includes chapters on the fabrication and study of various properties of ferrites. They discuss the various applications of ferrites in different areas as well as current and future challenges. All contributing authors are experts and active in their respective research areas. Proper references are provided at the end of each chapter, which direct readers to the best literature sources and help them obtain more details about the topic.

I am grateful to all authors for their help in completing this project. I am also thankful to the staff at IntechOpen for making this project possible. Furthermore, I am thankful to the Publishing Process Manager Mr. Kristian Josic for his cooperation throughout the publishing process. I hope that this book will advance and strengthen knowledge about ferrites and their applications in different areas.

Maaz Khan
Pakistan Institute of Nuclear Science and Technology,
Islamabad, Pakistan

Chapter 1

Synthesis, Characterization and Performance Evaluation of Magnetic Nanostructured $CoFe_2O_4$ for Adsorption Removal of Contaminant Heavy Metal Ions

Sumayya M. Ansari, Vikas Kashid, Bhavesh B. Sinha, Debasis Sen, Yesh D. Kolekar and Chintalapalle V. Ramana

Abstract

Engineering magnetic cobalt ferrite (CFO) nanomaterials for environmental remediation is difficult due to regeneration (without scarifying the magnetic properties), morphology with controlled size and shape, large-scale production, and thermochemical stability. Water management globally has struggled to remove hazardous heavy metals from water environments. We show an efficient, cost-effective, and low-temperature way to make highly nanocrystalline, regenerated inverse spinel CFO nanoparticles (NPs) and nanostructured CFO microgranules with improved magnetic properties that could be used to remove heavy metal ions (Pb^{+2}) from aqueous solutions without harming the environment. Magnetic investigations for CFO NPs reveal a saturation magnetization (M_S) of 3.09 μ_B/F.U. at 10 K, close to the expected value of a perfect inverted CFO structure (3.00 μ_B/F.U.). For CFO microgranules, the M_S is 5.62 μ_B/F.U. at 10 K, which is much higher than the bulk counterpart and nearly twice that of CFO NPs. Adsorption studies show that both magnetic adsorbents adsorb Pb^{+2} ions through a multilayer mechanism, as critically analyzed under the pseudo-first-order, pseudo-second-order, Elovich, Bangham's pore diffusion, and intraparticle diffusion models. CFO NPs and nanostructured CFO microgranules achieved 97.76% and 77.02% clearance efficiency, respectively.

Keywords: cobalt ferrite nanoparticles, spray drying, hydrothermal, nanostructured microgranules, heavy metal ions

1. Introduction

Inverse spinel cobalt ferrite ($CoFe_2O_4$) NPs feature extraordinary cubic magnetocrystalline anisotropy ($1.8-3.0 \times 10^6$ erg/cm^3) and tunable electrical characteristics, making them the focus of ongoing scientific investigation and technological applications [1, 2]. The $CoFe_2O_4$ NPs are studied experimentally for synthesizing, characterization, and applications in biomedical, electronics, memory devices, catalysis, high-performance microwave absorbers, and magnetic resonance imaging studies [3–10]. Magnetic $CoFe_2O_4$ NPS has currently grabbed the attention of the scientific and research community for its beneficial applications in environmental protection, particularly for contaminant and heavy metal ion adsorption [11–14]. Globalization, fast industrialization, urbanization, and population growth have polluted water, air, and soil. Drinking clean water is the most practical issue. Most chemical, electronics, and energy/power companies generate wastewater with hazardous metal ions. Heavy metal ions are persistent water pollutants [15–17]. Water pollution with hazardous metal ions (Cr^{3+}, Ni^{2+}, Co^{2+}, Cu^{2+}, Cd^{2+}, Ag^{2+}, Hg^{2+}, Pb^{2+}, and As^{2+}) is a major environmental and public health problem [18]. Heavy metals accumulate in the environment and cause heavy-metal toxicity. Thus, chemical, physical, and biological techniques have been devised to reduce pollution [19]. Among these processes, adsorption is one of the most widely used chemical processes for removing heavy metal ions and is considered easy to operate and cost-effective [15]. Until now, many adsorbents have been used to remove heavy metal ions, [20–22] and hence, the synthesis of novel adsorbents is of great interest in water treatment technology. These adsorbents are typically made of highly porous substances that provide the required surface area for adsorption [23]. Ideal adsorbent characteristics include a strong affinity for the target and a large surface area that provides numerous adsorption sites. Adsorbents should also be highly hydrothermally stable and highly resilient to severe conditions [24]. Using magnetic nanoparticles (MNPs) as adsorbents is an attractive option for overcoming the technical challenges for the reasons outlined below: Magnetic separation is regarded as a rapid, simple, and effective method for separating magnetic particles [25–28]. It has been used for mining ores, analytical chemistry, and biology. As adsorbents, various magnetic materials may be used [21, 25–28]. Due to their high chemical stability and modest saturation magnetization, MNPs of CFO with a cubic spinel structure have been created and used for contaminant adsorption [29]. For instance, Li et al. [30] demonstrated that the functional magnetic graphene sheets with CFO may adsorb methyl orange. Ai et al. [31] created composites out of activated carbon and CFO to remove the malachite green color from wastewater. In addition, Farghali et al. [32] prepared CFO/CNT composites for the removal of methyl green dye from aqueous solutions; however, the material displayed relatively poor adsorption capacity, perhaps as a result of the aggregation of CFO NPs on the surfaces of CNTs and the poor interactions between the CNTs and the NPs. More tweaking is required for magnetic materials to be more effective as absorbents. Such adjustments are made to create low-cost biosorbents that are amenable to large-scale pollution removal [33]. To the best of our knowledge, the preparation of magnetic CFO NPs, carbon-activated CFO composites, and surface-functionalized CFO NPs have been reported in the literature, and many coworkers have studied their dye and heavy metal ion removal from water [6, 20, 25–28, 30–32]. However, the quest for more sustainable, less time-consuming, and reproducible methods for large-scale synthesis is still being pursued. In this context, we have recently reported a simple one-pot synthesis of magnetic nanostructured CFO granules *via* spray drying [34]. Spray drying as a method of processing is considered environmentally friendly due to its utilization of low-cost reagents and aqueous conditions. Additionally, it is easily scalable to industrial applications. Thus, we have prepared CFO of different sizes as CFO NPs (CF1) and nanostructured CFO microgranules (CF2) and characterized both samples under identical conditions. Furthermore, we address the adsorption isotherms and kinetics studies of CF1 and CF2 to separate heavy metal ions.

2. Experimental

2.1 Sample preparation

The CFO NPs and CFO microgranules were prepared using hydrothermal [35], and spray drying processes [34]. During the preparation, stoichiometric molar amounts of $Co(NO_3)_2{\cdot}6H_2O$ and $Fe(NO_3)_3{\cdot}9H_2O$ were added into DI water and stirred well. Then pH of the solution was adjusted to 12 by adding ammonia solution (25%) and a homogeneous colloidal suspension was obtained at room temperature. For the synthesis of CFO NPs, the colloidal suspension was treated under hydrothermal conditions at 180°C for 24 h. The prepared particles were separated by centrifuge in the final solution. Finally, black precipitates were dried in an oven at 100°C overnight and designated as CF1. The colloidal solution, however, was made in the manner described above and then spray dried in a laboratory spray drier (model LU-228-Labultima; Mumbai, India). A compressed air spray nozzle created droplets between 10 and 20 μm. The aspiration flow rate was 45 m^3/h, and the input temperature was controlled at 170°C. The feed pump flow rate was controlled at 2 ml/min, and the atomization pressure was regulated between 2 and 2.5 kg/cm^2. As a further step, a glass cyclone separator was used to gather the spray-dried powder. The spray-dried powder was dark and free-flowing. The powder was then heated overnight at 400°C to produce CF2, a spray-dried CFO powder.

2.2 Characterization

Scanning Electron Microscopy (JEOL-JSM-6360) was utilized for morphology investigations. The CF1 and CF2 powder samples were subjected to Transmission Electron Microscopy (TEM) analyses using an FEI-Technai-G2-F30 microscope equipped with a Schottky field emission gun. The powder size distribution was estimated through micrograph image analysis utilizing Image-J software. A laboratory-based facility was utilized to conduct Small Angle X-ray Scattering (SAXS) experiments. The experiment involved the recording of scattered intensities I(q) as a function of the scattering vector transfer ($q = (4\pi \sin\theta)/\lambda)$), where "2θ" represents the scattering angle and "λ" represents the X-ray wavelength (λ = 0.154 nm) [36, 37]. The distance between the sample and detector was maintained at approximately 1070 mm. The powder samples' X-ray diffraction (XRD) patterns were obtained using a Bruker-D8-ADVANCE diffractometer. Determining lattice parameters was conducted through the Rietveld refinement methodology utilizing the FULLPROF SUITE software. After this, magnetic measurements were conducted utilizing the Quantum Design Evercool II PPMS-6000 apparatus, whereby magnetic fields were incrementally applied up to 90 ± kOe at both 10 K and 300 K. The study conducted low-pressure volumetric nitrogen adsorption-desorption measurements at a temperature of 77 K, which was maintained by a low-temperature liquid nitrogen bath. The measurements were carried out using an Autosorb iQ (Quantachrome Inc., USA) gas sorption system, with pressure levels ranging from 0 to 760 torr. Outgassing was executed under dynamic vacuum conditions (10–3 Torr) for 15 hours at a temperature of 200°C until a stable weight was attained. The study employed N2 of ultrahigh purity grade (99.999%), which underwent additional purification by utilizing calcium aluminosilicate adsorbents to eliminate minute quantities of water and other impurities before conducting the measurements. Ultra-pure helium gas (99.999% purity) was utilized to conduct warm and cold free-space correction measurements for N2 isotherms. About 200 mg of samples were used for the test, and their weight was recorded before and after outgassing to ensure that all moisture had been removed.

2.3 Adsorption experiments

2.3.1 Adsorption kinetic studies

$Pb(NO_3)_2$ was dissolved in DI water to make 20 mg/L Pb^{+2} aqueous solutions. Following that, investigations into adsorption involved combining 20 mg of magnetic adsorbents with 50 mL of heavy metal ion solutions in an aqueous medium. The pH of the solution was modified using standardized solutions of 0.1 M NaOH and 0.1 M HCl. The dispersions obtained were subjected to magnetic stirring at room temperature, and the temporal impact was assessed over a range of time intervals spanning from 5 to 300 min (specifically, 5, 10, 15, 30, 60, 90, 120, 150, 180, 210, 240, 270, and 300 min). A volume of 2 mL of solution was obtained, and the magnetic adsorbents were eliminated through the process of magnetic separation. The quantification of the Pb + 2 ion concentration was performed using atomic absorption spectroscopy (AAS) with a Varian Spectr AA-220 instrument. Eqs. (1)and (2) were utilized to compute the quantities of metal ions adsorbed per unit mass of the adsorbent and the corresponding removal efficiencies (R).

$$q_e = \frac{V(C_i - C_e)}{1000 \times S} \tag{1}$$

$$R = \left(\frac{C_i - C_e}{C_i}\right) \times 100 \tag{2}$$

where C_i and C_e are the concentrations (mg/L) of the metal ions in the aqueous solution before and after the adsorption period, respectively. V denotes the volume (mL) of the aqueous solution, and S represents the amount of dry adsorbent used (g).

2.3.2 Adsorption isotherm studies

The present study investigated adsorption isotherm to examine the equilibrium relationship between adsorbents and adsorbates. The study involved the acquisition of adsorption isotherms of Pb^{+2} on magnetic adsorbents, and this was achieved through the dispersion of 20 mg of magnetic adsorbent into 30 mL of Pb^{+2} ion solution, with varying concentrations between 20 and 1000 mg/L at ambient temperature. The dispersions were subjected to magnetic stirring under ambient conditions, and a volume of 2 mL of the resultant solution was extracted after a duration of 30 minutes. The magnetic adsorbents were extracted through magnetic separation, and the concentration of heavy metal ions was measured using atomic absorption spectroscopy.

2.3.3 Recovery and reuse

The magnetic adsorbents, loaded with Pb^{+2} (20 mg), were subjected to stirring with a 0.1 M HCl solution (10 mL) at room temperature for a duration of 3 h to facilitate desorption of the metal ions. The concentration of the metal ion in the aqueous phase was determined using AAS. Subsequently, the magnetic nanoparticles (MNPs) were subjected to neutralization using a diluted solution of 0.1 M NaOH, followed by a thorough rinse with deionized water. The colloidal magnetic adsorbents were subsequently extracted through magnetic separation. The MNPs were then subjected to further adsorption processes to assess their reusability. The magnetic adsorbents were subjected to 5 cycles of adsorption and desorption.

3. Results and discussion

3.1 Morphology, microstructure, and crystal structure

The surface morphology of pristine CF1 and CF2 samples is shown in **Figure 1(a, b)**. Nearly spherical morphology with mean particle size (D_{mean}) ~19.84 nm and 111 nm, for CF1 and CF2, respectively, is evident from **Figure 1**. The CF2 microgranules were made up of subunit entities. When the dimension is reduced to the nanoscale, two effects are expected: (i) a significant increase in relative

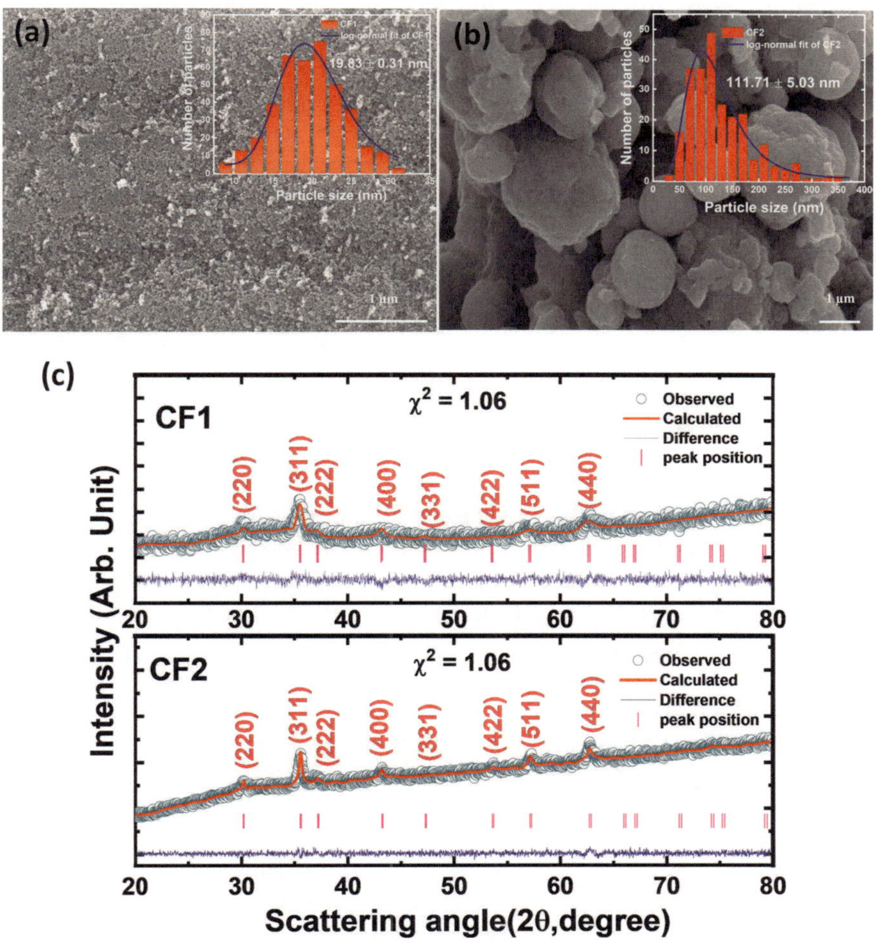

Figure 1.
Morphological study of CF1 (a) and CF2 (b) (inset shows particle size distribution), and Rietveld refined XRD patterns for CF1 and CF2 samples (c).

surface area and (ii) a significant increase in the number of atoms on the surface [38]. These effects enhance the chemical reactivity of the surface atoms of nanoparticles. Therefore, at the nanoscale, the Vander Waals forces of attraction as well as the magnetic force of attraction tend to increase, and subunit entities tend to attract each other, resulting in the assembly of subunit particles in a regular or irregular manner [38]. The size distribution (**Figure 1(c,d)**) for CF1 is observed to be symmetrical with small polydispersity index ($\sigma \leq 0.25$) compared to CF2 ($\sigma = 0.43$), indicating that the CF1 NPs are well confined to a limited diameter range which is desirable for practical applications. **Figure 1(c)** shows the differences between calculated and observed XRD patterns as well as the crystalline phases of CF1 and CF2. A Rietveld refinement was carried out until the fitment produced a goodness factor (χ^2) $\sim$ 1. The Rietveld refinement of XRD data confirms the single-phase cubic spinel structure of both samples, which are devoid of impurities. The obtained refinement parameters, the discrepancy factor (R_{wp}), expected values (R_{exp}), and χ^2 for CF1 and CF2, are (76.5, 66.20), (74.34, 64.31), and 1.06, in good accord with the reported parameters for ferrite systems [1, 34]. The presence of intense diffraction peaks indicates that both samples are crystalline. All peaks were indexed based on the structure of cubic spinel ferrite (JCPDS card number 22-1086) with a space group of Fd3m (2 2 7). However, the diffraction peaks of CF1 are relatively broad, which can be attributed to its reduced crystallite size. The average crystallite size was estimated using the integral width of the diffraction lines, Scherrer's formula, background subtraction, and correction for instrumental broadening; For CF1 and CF2, the estimated values are $\sim$20.05 $\pm$ 0.05 nm and $\sim$ 26.41 $\pm$ 0.05 nm, respectively. The lattice parameters for CF1 (8.385 Å) and CF2 (8.373 Å) differ marginally from one another and deviate slightly from those of cobalt ferrite in bulk (8.394 Å). This could be the result of nanosize effects [1, 34]. The calculated X-ray density for CF1 and CF2 is 5.509 g/cm^3 and 5.314 g/cm^3, respectively. Refinement of experimental data for CF1 demonstrates an ideal inverse cation distribution as $[Fe^{+3}]_{Tet}\{Co^{+2}Fe^{+3}\}_{Oct}O^{-2}{}_4$, whereas CF2 deviates from this ideal by transferring some Co^{+2} cations to the tetrahedral A-site, and cation distribution for CF2 is, that is, $[Co^{+2}{}_{(0.6875)}Fe^{+3}{}_{(0.3125)}]_{Tet}\{Co^{+2}{}_{(0.3125)}Fe^{+3}{}_{(1.6875)}\}_{Oct}$ $O^{-2}{}_4$. A detailed study of the structural and magnetic properties of both samples was made by first-principles calculations and was reported in our earlier work [29].

The TEM micrographs of CF1 and CF2 samples are shown in **Figure 2(a-d)**. CF1 results in a spherical shape with some cubic-like morphology, as shown in **Figure 2a**. The average size of particles is $\sim$16.11 nm (shown by the red circle and cube). The high magnification image (**Figure 2b**) shows the CF1 samples are composed of small NPs with a spherical shape. Most NPs have a size smaller than 10 nm (range of 5.0–6.5 nm) (see inset of **Figure 2b**). These NPs are self-assembled in a spherical, close-packed super-lattice due to the high degree of uniformity in diameter. On the other hand, CF2 shows a quasi-spherical shape with cube morphology (**Figure 2(c,d)**). The average size is 22 nm, and the magnified image shows the dominance of cube morphology for CF2 (**Figure 2d**). The electron diffraction patterns shown in **Figure 2(e,f)** obtained through the selected area electron diffraction (SAED) technique exhibit diffuse rings that can be attributed to the (220), (311), (400), (511), and (440) crystallographic planes of the $CoFe_2O_4$ cubic structure. The manifestation of CFO's polycrystalline character is apparent through the existence of numerous diffraction rings in the corresponding SAED patterns for both specimens. Note that the particle boundary is well defined for both the samples and an isolated cube shows (inset of **Figure 2c**) more clearly a size of 16.77 nm. **Figure 3(a–f)** show HRTEM images of CF1 and CF2 samples, respectively. The clear lattice boundary in the HRTEM image

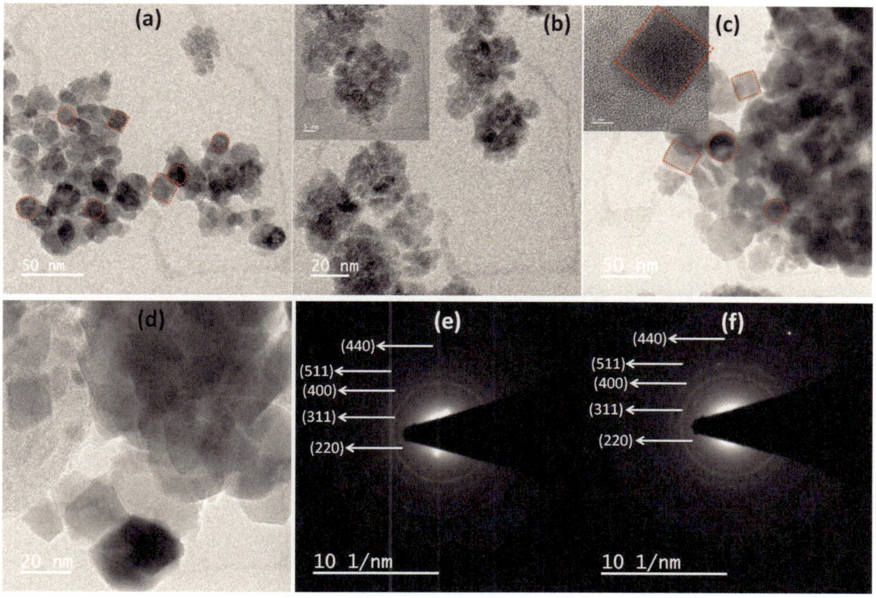

Figure 2.
TEM images of CF1 (a, b) and CF2 (c, d) along with a magnified view (inset) with a scale bar of 5 nm and SAED patterns of CF1 (e) and CF2 (f) samples.

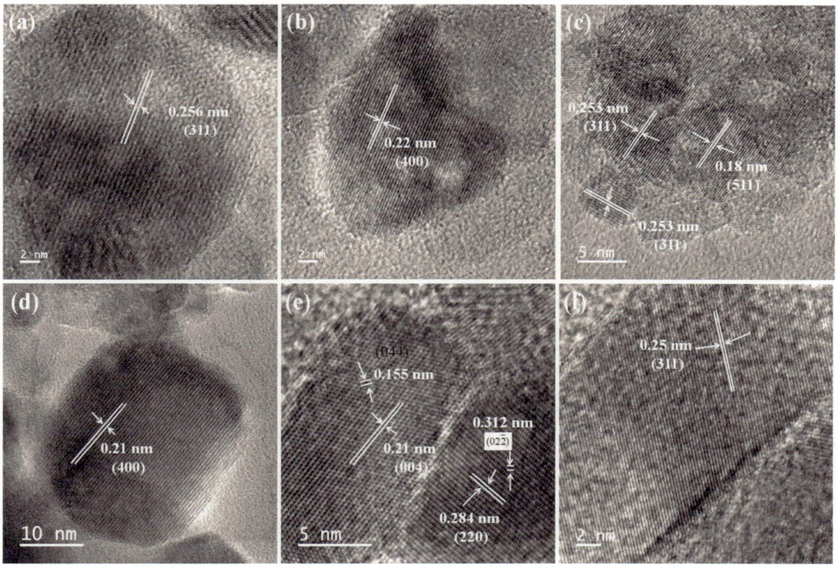

Figure 3.
HR-TEM images of CF1 (a–c) and CF2 (d–f).

illustrates the high crystallinity of both samples. The periodic fringe spacing of (0.253–0.2256 nm), 0.22 nm, and 0.18 nm corresponds to the (311), (400), and (511) planes of cubic $CoFe_2O_4$ as observed for CF1 (**Figure 3(a–c)**). The periodic fringe spacing of 0.21 nm, 0.155 nm, 0.312 nm, 0.284 nm, and 0.25 nm corresponds to the (400), $(\bar{0}44)$, (022), (220), and (311) planes of cubic $CoFe_2O_4$ as observed for CF2 (**Figure 3(d–f)**), which matches with JCPDS card no. 22-1086.

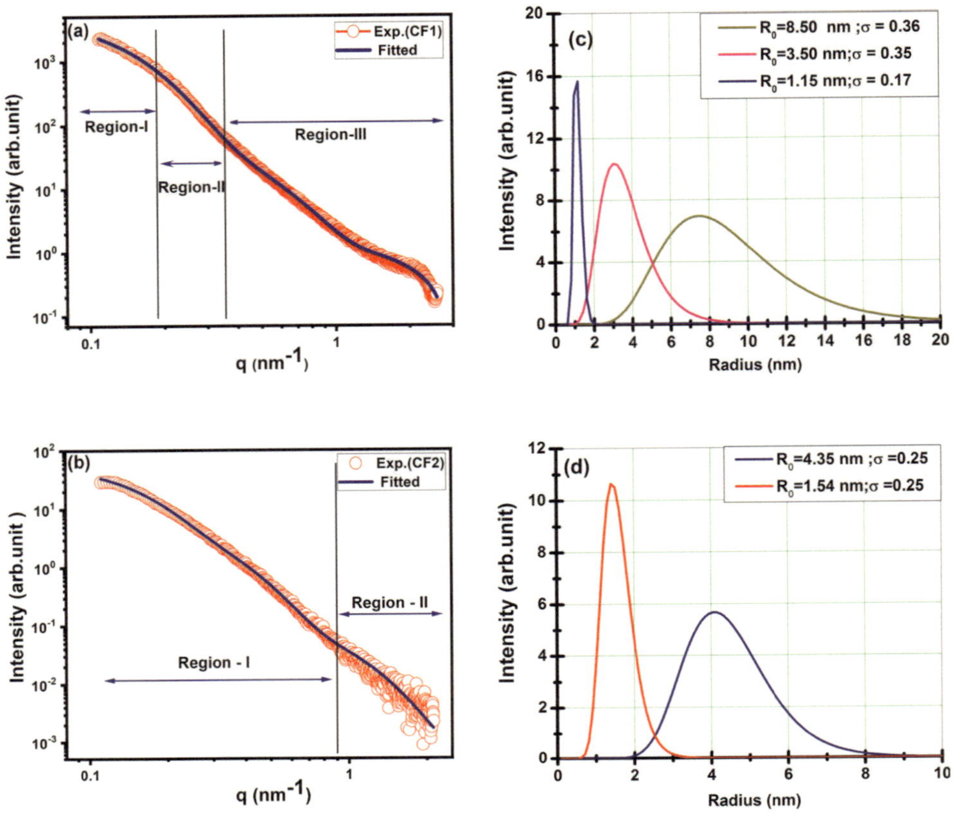

Figure 4.
SAXS profile for CF1 (a), and CF2 (b) and the calculated size distribution are shown in (c), and (d).

3.2 Size and confined structure

The basic size of particles, morphology, and corelation of interlocked nanostructure in terms of structure factor can be well understood using SAXS analysis.

The SAXS profile for CF1 and CF2 are shown in **Figure 4(a,b)**. It is observed that the scattering profiles could be best represented by the following contributions:

$$I_{CF1}(q) = \sum_{i=1}^{3} I_i.(q) \tag{3}$$

and

$$I_{CF2}(q) = \sum_{i=1}^{2} I_i.(q) \tag{4}$$

Here, "I_i" denote the contribution to scattering intensity from "i^{th}" component. For an ensemble of interacting spherical particles, under local monodisperse approximation, intensity can be written as: [37]

$$I_i(q) = C_i \int_0^{\infty} P_i(q,R)\, S_i(q,R)\, R_i^6 D_i(R)\, dR \tag{5}$$

Here, C_i is a scale factor, $P_i(q, R)$ and S_i (q, R) signify the form factor and interparticle structure factor for the "i^{th}" component, respectively. Assuming the spherical shape of particles with radius R, the form factor P (q, R) is expressed as [39]:

$$P_i(q,R) = \left[\frac{3\sin(qR) - qR\cos(qR)}{(qR)^3}\right]^2 \tag{6}$$

$D_i(R)$ represents the size distribution of "i^{th}" component and is assumed to be a normalized log-normal distribution: [40]

$$D_i(R) = \frac{N_i}{\sqrt{2\pi\sigma^2}R}\, exp\left[-\left[ln\left(R/R_0\right)\right]^2/2\sigma^2\right] \tag{7}$$

where N_i represents the normalization factor, R_0 and σ ($0 < \sigma < 1$) represent the median radius and polydispersity index for the ith distribution, respectively. As the magnetic attraction among the particles play a significant role, consideration of an attractive potential was necessary during the detailed fitting procedure and mass fractal type structure factor was considered for $I_3(q)$ and $I_2(q)$ contributions for CF1 and CF2, respectively. In such a case, S(q) can be represented as:

$$S_i(q) = 1 + \frac{D}{r_0^D}\int_0^\infty r^{D-3}h(r,\xi)\frac{\sin(qr)}{qr}r^2dr \tag{8}$$

$$\text{with cut} - \text{off function as}: h(r,\xi,\alpha) = exp\left[-\left(\frac{r}{\xi}\right)^\alpha\right] \tag{9}$$

where r-dimension of individual scattering objects, ξ (x_i)- size of aggregate or cut-off length for the fractal correlations, and D-fractal dimension ($1 < D < 3$). The model scattering intensity, as mentioned above using Eqs. (3) and (4), was fitted to the scattering data for CF1 and CF2 using the nonlinear least square method. It was found that the mass fractal model [41] could describe the present scattering profiles when compared to the hard-sphere and sticky hard-sphere model [42]. For an attractive potential, the intensity at low q increases because of the formation of aggregated structure. **Figure 4(c,d)** shows the size distribution obtained from SAXS profile for the CF1 and CF2 samples while the parameters derived are summarized in **Table 1**. It is worth mentioning that the real space and scattering space are connected by Fourier transform; thus, the region-I, II for CF1 and region-I for CF2 of the scattering profile primarily contain information about large length scales. Similarly, region-III for CF1 and region-II for CF2 contains information about the structure and correlation at a

Sample	Region	R_0 (nm)	σ (nm)	r_0 (nm)	x_i (nm)	D	A
CF1	I	8.33	0.20	1.37	4.50	2.24	2.95
	II	3.03	0.25				
	III	1.15	0.17				
CF2	I	4.35	0.25	3.98	15.15	2.25	2.27
	II	1.54	0.25				

Table 1.
Structural parameters obtained from small angle scattering experiments.

smaller length scale (i.e., correlated nanoparticles and nano-meso-pores, etc.). In case of CF1, the scattering model suggests that the CF1 NPs are composed of spherical shape particles of median size $R_o \sim 8.33$ nm with polydispersity index (σ) ~ 0.20 (region I). This is further evident in **Figure 2(a)**, that is, TEM images of CF1. Furthermore, region II and III shows the presence of spherical shape particles of Ro ~ 3.03 nm and 1.15 nm with $\sigma \sim 0.25$ and 0.17, respectively. The values agree with TEM data. The smaller value of σ manifests the monodisperse size behavior of NPs. Furthermore, structure factor analysis shows that the NPs are composed of monomer with radius ~ 1.37 nm, fractal dimension (D) is found to be ~2.24, which results in a maximum size of ~4.50 nm. It is more clearly seen in the inset of image **Figure 2b**. In the case of CF2, as discussed above, the high magnetization value of nanostructured CFO develops strong magnetic attraction that results in stable microgranules. For CF2, the scattering model fitted using Eq. (7) suggests that the basic CF2 sample is composed of particles of mean radius ~4.35 nm with polydispersity index ~0.25. Note that region I cover 80% of whole *q* range and simultaneously, the interparticle structure factor suggests the maximum aggregate size ~15.15 nm as seen in TEM analysis. The remaining 20% of *q* range suggests a mean diameter of 3.18 nm. However, the mean size of microgranules as seen from SEM micrographs is $D_{mean} \sim 111$ nm and this suggests that the limitation of accessible low "*q*" range in SAXS. However, SAXS profile strongly suggests that the CF2 sample is nanostructured microgranules. Moreover, it was observed that characteristic dimension of individual scattering objects (r_0) and maximum aggregate size (x_i) value is smaller ($r_0 \sim 1.37$, $x_i = 4.50$) for CF1 compared to CF2 ($r_0 \sim 3.98$, $x_i = 15.15$). Note that the size obtained from TEM is well agreed with the SAXS analysis.

3.3 Magnetism

Figure 5 shows the experimental magnetization (M-H) loops for CF1 and CF2 measured at 10 K and 300 K temperatures, and the magnetic parameters obtained are enlisted in **Table 2**. The observed value of M_S at 10 K is ~73.69 emu/g (3.09 μ_B/F.U.) and 133.79 emu/g (5.62 μ_B/F.U.) for CF1 and CF2, respectively. Considering the formula $[Co^{+2}_{(1-x)}Fe^{+3}_{(x)}]_{Tet}\{Co^{+2}_{(x)}Fe^{+3}_{(2-x)}\}_{Oct}O^{-2}_4$ to describe the cation distribution in the spinel structure of CFO and assuming that Fe^{3+} and Co^{2+} ions have magnetic moment of 5 μ_B and 3 μ_B, respectively, and the inversion parameter is obtained to be $\delta \sim 0.97$ and 0.345 for CF1 and CF2, respectively. This value of inversion parameter indicates that the crystal structure of CF1 is very close to the inverse spinel and non-stoichiometric inverse spinel behavior for CF2 sample. Notice that, for CF1, the experimental M_S value is close to the theoretical value of ideal inverse CFO structure (~3.00 μ_B/F.U.). For CF2, M_S is remarkably high compared to the bulk counterpart, and almost double as compared to M_S value reported for CFO NPs [2, 3]. This remarkable increased M_S value for CF2 may be attributed to the non-stoichiometric cations (Fe^{+3}, Co^{+2}) distribution among the octahedral and tetrahedral sites as compared to ideal spinel structure $[Fe^{+3}]\{Co^{+2}Fe^{+3}\}O^{-2}$ as predicted by the XRD analyses.

3.4 BET analysis

Notably, gas absorption (BET) techniques are appropriate for probing surface areas in porous materials. As illustrated in **Figure 6**, N_2 adsorption-desorption isotherms were measured to ascertain the absorptive capacity of magnetic adsorbents for gas absorption. According to the IUPAC classification observed for CF1 and CF2, the

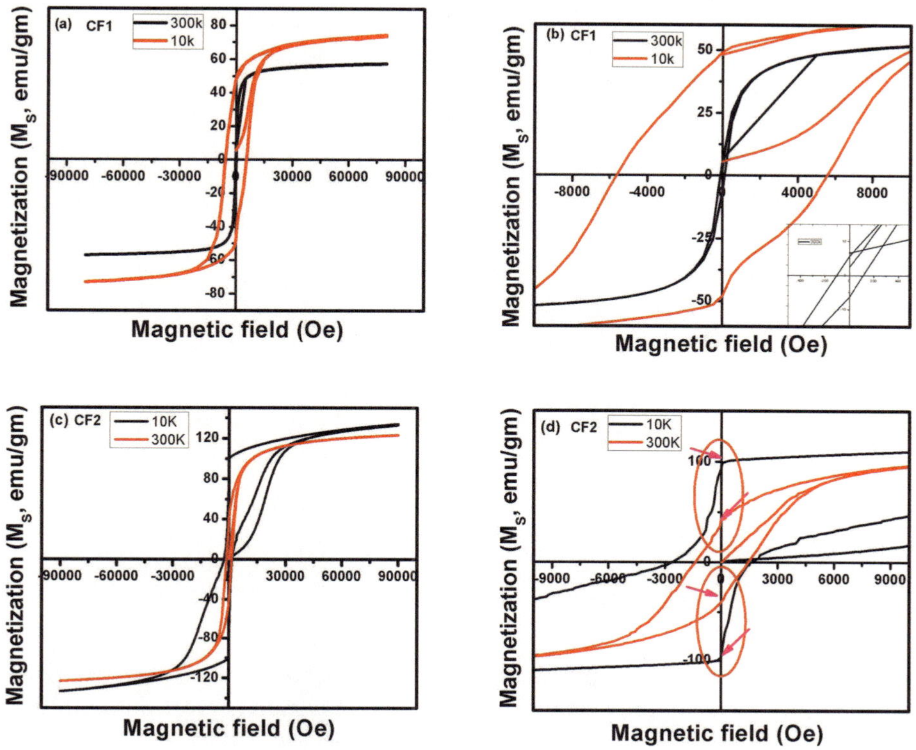

Figure 5.
M-H curve for CF1 (a, b) and CF2 (c, d) at 10 K and 300 K.

Sample	Temperature (K)	Saturation magnetization (M_S) (emu/gm)	Saturation magnetization (M_S) (μ_B/F.U.)	Remanent magnetization (M_r) (emu/g)	Coercivity (H_C) (Oe)	Squarness ratio Mr./Ms
CF1	300	57.313 ± 0.287	2.407 ± 0.012	5.966 ± 0.030	109 (± 0.109)	0.104
	10	73.687 ± 0.368	3.095 ± 0.015	48.608 ± 0.243	5575 (± 5.575)	0.660
CF2	300	123.09 ± 0.615	5.170 ± 0.026	40.450 ± 0.202	1410 (± 1.410)	0.329
	10	133.79 ± 0.669	5.620 ± 0.028	96.50 ± 0.483	2118 (± 2.118)	0.721

Table 2.
Magnetic parameters obtained from M-H measurements at 10 K and 300 K.

N2 gas adsorption-desorption isotherm exhibits a type IV curve and an H3 hysteresis loop. This behavior indicates that mesopores predominate [43]. The hysteresis of type H3 reveals the random distribution and interconnection of pores. Because adsorption and desorption isotherms exhibit distinct behaviors to the pore network at a relative pressure of 0.45 (for N_2 at 77 K), these pore properties significantly influence the desorption isotherm more than the adsorption isotherm. A BET surface area measurement was performed to ascertain the prepared material's surface area. Using the BET multipoint method, the specific surface area of CF1 and CF2 was determined to be

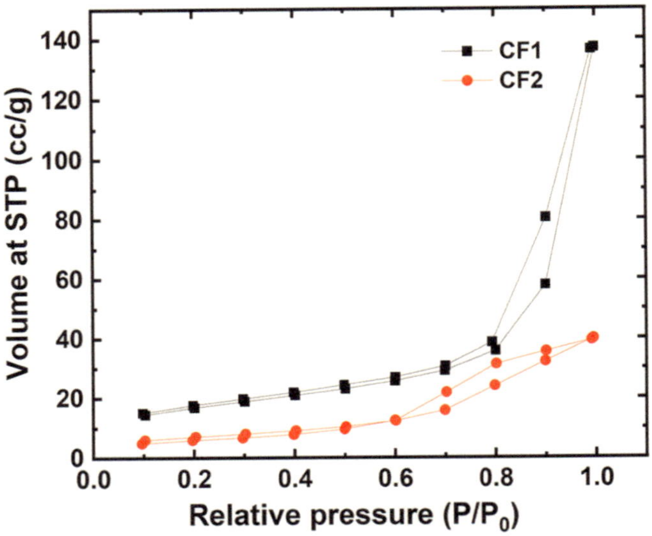

Figure 6.
BET hysteresis curve during adsorption and desorption for CF1 and CF2.

57.66 m^2/g and 24.67 m^2/g, respectively. Thus, both magnetic adsorbents are porous, and it is noteworthy that the average pore size of CF1 is more significant than that of CF2 (7.347 nm vs. 4.994 nm). It is evident that the specific surface area, pore availability, and affinity between the adsorbate and adsorbent significantly influence the adsorption capacity, which indicates the presence of active sites for the absorption of additional Pb^{+2} ions.

3.5 Adsorption studies

3.5.1 Adsorption kinetics studies

Figure 7(a) shows the effect of time on the Pb^{+2} ions concentration at room temperature (RT) during adsorption experiments and it is seen that the Pb^{+2} concentration decreases with increasing time for the magnetic adsorbents. Although, Pb^{+2} concentration decreases with a relatively slow rate for CF2 compared to CF1. However, Pb^{+2} concentration decreases rapidly, up to 3.70 mg/L within 90 mints compared to initial Pb^{+2} concentration (i.e., 20 mg/L) when CF1 was used for adsorption. Whereas for CF2 a rapid decrease of Pb^{+2} concentrations up to 13.40 mg/L was observed within 30 mints compared to initial Pb^{+2} concentration as 20 mg/L. **Figure 7(b)** shows the effect of time on the adsorption capacity of Pb^{+2} at RT. In case of nano-adsorbent, at the beginning (up to 90 min), the rate of adsorption is relatively fast and further the rate increases gradually and finally slows down to attain equilibrium indicating a decrease in the number of available sites as the adsorption proceeds. On the other hand, for CF2 the rate of adsorption was observed to be fast up to 30 mints and the adsorption capacity increases from 16.51 to 35.76 mg/g with a relatively slow rate and then attains equilibrium. Moreover, the adsorption process reaches equilibrium within 210 mints and 240 mints for CF1 and CF2, respectively. **Figure 7 (c)** depicts the time-dependent removal efficiency of Pb^{+2} ions. Here, 50% of the Pb^{+2} ions were completely absorbed in the first 30 min by CF1 compared to the initial

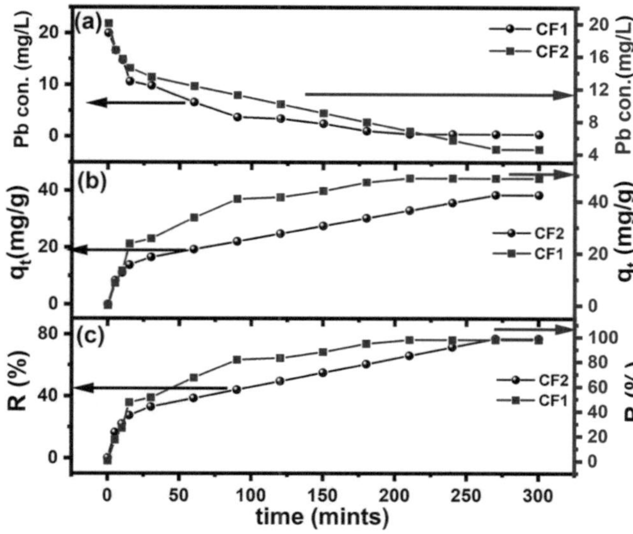

Figure 7.
Effect of time on the Pb concentration (mg/L) (a), Adsorption capacity (q_t) (b) and removal efficiency (R %) (c), during adsorption process for magnetic adsorbents.

concentration (20 mg/L) of Pb^{+2} as the removal efficiency was observed to be 51.09%; whereas CF2 attains 49.51% removal efficiency in 270 min. It is noticeable that the maximum removal efficiency was observed to be 97.76% and 77.02% for CF1 and CF2, respectively. And, the maximum adsorption capacity ($q_{e,exp}$) was observed to be 48.88 mg/g and 38.51 mg/g for CF1 and CF2, respectively. It demonstrates that the adsorption capacity of CF1 is slightly more compared to CF2. This more adsorption observed for CF1 could be attributed to the larger specific surface area of porous adsorbents that offers the higher surface energy for adsorbing heavy metal ions. Since adsorption is particle diffusion controlled, an increase in pore increases the number of accessible sites, hence increases the amount of adsorbate (Pb^{+2}) on the adsorbent [44–46]. The effect of particle size on adsorption of the metal ions from aqueous solutions has been reported [46] and our results are well agreed.

To understand the detailed adsorption mechanism, its kinetics are analyzed by a few models based on the adsorption equilibrium. The experimental data were fitted to the pseudo-first-order [47], pseudo-second-order [47], intraparticle diffusion, Bangham's pore diffusion, Boyd kinetic model, and Elovich models; these equations are shown in **Table 3**.

Figure 8(a,b) shows the pseudo-first- and second-order kinetic model plot for CF1 and CF2, from which k_1, k_2, and $q_{e,cal}$ values are extracted and enlisted in **Table 4**. It has been observed that for both the adsorbents, the correlation coefficient (R^2) calculated using pseudo-second-order model, was found to be larger (0.9948 ± 0.0058 and 0.9413 ± 0.0058) than those observed for pseudo-first-order model (0.9541 ± 0.145 and 0.9192 ± 0.145). The experimental ($q_{e,exp}$) and calculated ($q_{e,cal}$) values of q_e using pseudo-first-order model do not match with each other, whereas the values are in good agreement with each other using pseudo-second-order for CF1. On the other hand, the $q_{e,exp}$ and $q_{e,cal}$ values using pseudo-first-order and second-order-model match with each other for CF2 absorbent. Thus, based on R^2 value, we can predict that the overall adsorption of Pb^{2+} onto magnetic adsorbents followed the

Model	Linear equation	Plot	Calculated coefficient
Pseudo-first-order [a]	$log\,(q_e - q_t) = log\,(q_e) - \frac{k_1 t}{2.303}$	$log\,(q_e - q_t)$ vs. (t)	$k_1 = -slope \times 2.030$ $q_e = e^{intercept}$
Pseudo-second-order [b]	$\left(\frac{t}{q_t}\right) = (t)\left(\frac{1}{q_e}\right) + \left(\frac{1}{k_2 q_e^2}\right)$	(t/q_t) vs. (t)	K_2 = slope^2/intercept q_e = slope^{-1}
Intraparticle diffusion [c]	$q_t = k_i t^{0.5} + C$	(q_t) vs. $(t^{0.5})$	K_i = slope C = intercept
Bangham's pore diffusion[d]	$log\left[log\left(\frac{C_i}{C_i - mq_t}\right)\right] = log\left(\frac{K_b m}{2.303V}\right) + \alpha log(t)$	$log\,[log(Ci/Ci\text{-}mq_t)]$ vs. $log(t)$	$\alpha = slope$ $K_b = e^{intercept} \times 2.030 \times V/m$
Boyd kinetic[e]	$F = 1 - \frac{6}{\pi^2} exp\,(-B_t)$ $B_t = -0.4977 - ln\,(1 - F)$	—	—
Elovich model[f]	$q_t = \beta ln(\alpha\beta) + \beta(t)$	q_t vs. $ln(t)$	$\beta = slope$ $Intercept = \beta\, ln(\alpha\beta)$

[a, b]Where $k_1(mint)^{-1}$, k_2 $(g.mg^{-1}mint^{-1})$ are the pseudo-first- and second-order rate constant and q_e (mg/g) and q_t (mg/g) are the adsorption capacity of Pb^{+2} onto adsorbent at equilibrium and at a given contact time t (min), respectively.

[c]$k_i(mg/g.min^{1/2})$ is the intraparticle diffusion rate constant and C(mg/g) is a constant that gives an idea about the thickness of the boundary layer.

[d]C_i is the initial adsorbate concentration in liquid phase, m is the weight of the adsorbent, q_t(mg/g) same as described above, V is volume of the solution, and α ($\alpha < 1$), K_b(L/gm) are constants.

[e]F is the fraction of solute adsorbed at different contact time t and parameter B_t is a mathematical function of F, which is given by F = q_t/q_e.[f]α(mg /g min) is the initial sorption rate and the parameter β (mg/g min$^{1/2}$) is related to the extent of surface coverage and activation energy for chemisorption.

Table 3.
Mathematical equations applied in the kinetic adsorption study of Pb^{+2} ions onto CF1 and CF2.

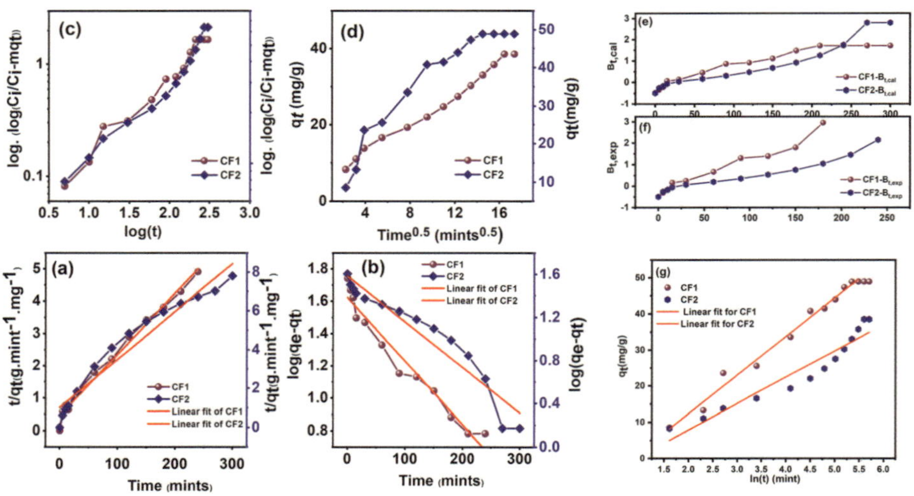

Figure 8.
Lagergren pseudo-first-order (a), Lagergren pseudo-second-order (b), Bangham's pore diffusion (c), intraparticle diffusion (d) model, and plots of calculated Boyd parameter ($B_{t,cal}$) versus time (t) (e), experimental Boyd parameter ($B_{t,exp}$) versus time (t) (f), and Elovich model plot and (g) for CF1 and CF2 adsorbents.

pseudo-second-order model. Moreover, for CF1, values of $q_{e,cal}$ and $q_{e,exp}$ were found to be larger (54.95 mg/g and 48.88 mg/g) than those observed for CF2 adsorbents (40.00 mg/g and 38.51 mg/g) as per the pseudo-second-order model.

Pseudo-first-order model

Sample	k_1 ($\times 10^{-3}$) (mint)$^{-1}$	$q_{e,cal}$ (mg/g)	R^2
CF1	8.98	41.69	0.9541
CF2	9.67	38.01	0.9192

Pseudo-second order model

Sample	$q_{e,exp}$ (mg/g)	k_2 ($\times 10^{-4}$) (g.mg^{-1}mint^{-1})	$q_{e,cal}$ (mg/g)	R^2
CF1	48.88	5.90	54.95	0.9948
CF2	38.51	6.18	40.0	0.9413

Intra-particle diffusion model

Sample	k_i (mg /g min$^{1/2}$)	C (mg/g)	R^2
CF1	2.94	7.85	0.9308
CF2	1.96	4.58	0.9891

Bangham's pore diffusion model

Sample	A	k_b (L/gm)	R^2
CF1	0.73	7.48	0.9779
CF2	0.49	9.50	0.9611

Elovich model

Sample	α (mg /g min)	β (mg/g min $^{1/2}$)	R^2
CF1	0.040	10.70	0.9847
CF2	0.125	7.30	0.9127

Table 4.
Characteristics of the pseudo-first-order, pseudo-second-order, intra-particle diffusion, Bangham's pore diffusion, and Elovich kinetics model, along with the correlation coefficient (R^2) for the adsorption of Pb^{+2} (initial concentration C_i = 20 mg/L) onto the CF1 and CF2 adsorbents.

From a mechanistic viewpoint, it is crucial to identify the steps involved during the adsorption process. Thus, the intraparticle diffusion model [43] has been used to identify the steps involved during adsorption process. **Figure 8(c)** shows the intraparticle diffusion model plot for CF1 and CF2 and here, the non-zero value of *C* implies that the adsorption mechanism is governed by both film diffusion and intra-particle diffusion [47, 48]. In the experiment for CF1, the plot shows three main portions: initial linear but rapid increase (15%) portion, steep portion (70%), and a later horizontal potion (15%), whereas for CF2 adsorbents **Figure 8(c)** shows two distinct portion as first linear steep portion (90%) and later horizontal portion (10%). In the case of CF1, the initial rapid portion attributed to boundary layer diffusion effect or external mass transfer effect. The boundary layer effect is more dominant in CF1 compared to CF2 as *C* value is large (see **Table 5**). However, the steep linear portion implies that the intraparticle diffusion could also be the main sorption mechanism in this part of plot. The later portions were slow and controlled by the equilibrium diffusion mechanism, which occurred when the rate of sorption and desorption are insignificant. Moreover, 90% of the linear steep region observed for CF2 indicates that intraparticle diffusion is more dominant for CF2. In addition, the rate of adsorption process (k_i) of CF1 (2.935 ± 0.252 mg.g^{-1}mint$^{-0.5}$) is larger than the CF2 (1.096 ± 0.059 mg.g^{-1}mint$^{-0.5}$). The diffusion rates (**Table 5**) decrease with the

Model	Linear equation	Plot	Calculated coefficient
Langmuir[a]	$\left(\frac{C_e}{q_e}\right) = \left(\frac{1}{bq_m}\right) + \left(\frac{C_e}{q_m}\right)$	(C_e/q_e) vs. (C_e)	b = *slope/intercept* q_m = slope^{-1}
Freundlich[b]	$q_e = k_f C_e^{1/n}$	log (q_e) vs. log (C_e)	$k_f = 10^{\text{intercept}}$ n = slope^{-1}
Tempkin[c]	$q_e = k_1 \ln(k_2) + k_1 \ln(C_e)$	$ln(C_e)$ vs. (q_e)	k_1 = slope^{-1} $k_2 = 10^{\text{intercept}}$

[a]*$q_e(mg.g^{-1})$ is the amount of Pb^{+2} ions adsorbed per unit mass of magnetic adsorbent, C_e is the concentration of adsorbate in the solution at equilibrium, $q_m(mg.g^{-1})$ is the maximum uptake per unit mass of adsorbent or monolayer adsorption capacity, and $b(g.mg^{-1})$ is the Langmuir constant related to the adsorption energy.*[b]*where C_e and q_e have the same meaning as in the Langmuir isotherm, $k_f(mg.g^{-1})$ is the Freundlich constant related to the adsorption capacity (mg/g), and n is the empirical parameter representing the energetic heterogeneity of the adsorption sites (dimensionless).*[c]*$k_1(L/g)$ is related to the heat of adsorption (L/g), and k_2 is the dimensionless Tempkin isotherm constant.*

Table 5.
Mathematical equations applied in adsorption isotherm study of Pb^{+2} ions onto CF1 and CF2 samples.

increase in contact time for both adsorbents. This is because the heavy metal ions diffuse into the inner structure of the adsorbent and the pores for diffusion become smaller. The free path of the ions in the pore decreases and the ions may also be blocked. Here, the deviation of straight lines from the origin (**Figure 8(c)**) was observed for both the adsorbents. It may be due to the difference in rate of mass transfer in the initial and final stages of adsorption. Further, such deviation of straight line from the origin indicates that the pore diffusion is not the sole rate-controlling step.

The experimental data were further analyzed to determine the slow step occurring in the present system using Bangham's pore diffusion model [48] and Bangham's pore diffusion model plots are shown in **Figure 8(d)**. It is found to be linear with R^2 of $\sim$0.9779 and $\sim$ 0.9611 for CF1 and CF2, respectively, which confirm that the adsorption is pore-diffusion controlled for both the adsorbents. Pore diffusion is more dominant for nano-adsorbent compared to micro-adsorbent as pore presence were already seen in BET isotherm curve.

Additionally, the actual rate-controlling step involved in the adsorption process was determined by Boyd kinetic model [49]. Using the $q_{e,exp}$ and $q_{e,cal}$ values, the corresponding experimental Boyd ($B_{t,exp}$) and calculated Boyd ($B_{t,cal}$) parameters were obtained and as shown in **Figure 8(e,f)**, the plots are linear but do not pass through the origin, and the R^2 of linear fitting indicates that the external mass transport mainly governs the rate-limiting process of adsorption of Pb^{+2} onto the magnetic adsorbents and it is more prominent in magnetic nano-adsorbent [50].

Moreover, the data were further analyzed using the Elovich model [51] and the linear form of Elovich model is presented in **Table 3**. The unknown constants α and β were obtained and listed in **Table 4**. Elovich model plot is shown in **Figure 8(g)**. The kinetic curve of adsorption of Pb^{+2} was demonstrated to be good fitting with the Elovich model for CF1 and CF2 as the R^2 is observed to be near to unity. However, from **Table 4** it is seen that the diffusional rate-limiting is more prominent for the nano-adsorbent compared to nanostructured microgranules adsorbent during the adsorption process.

3.5.2 Adsorption isotherm studies

The Langmuir and Freundlich models were utilized to fit the experimental data, as presented in **Table 5**. These equations are widely employed in the analysis of adsorbate-adsorbent interactions. The Langmuir adsorption model [52] postulates the existence of a maximum capacity for adsorption, which corresponds to a state of complete saturation of the adsorbent surface by a monolayer of adsorbate molecules. It is commonly assumed that the process of adsorption occurs at distinct and uniform sites located within the absorbent material. Upon occupation of a site by Pb^{+2} ions, subsequent adsorption at said site is precluded. The determination of q_m and b values is achieved through the computation of the slope and intercept of the linear graph of (C_e/q_e) against (C_e), as illustrated in **Figure 9(a)**. The resulting parameters are presented in **Table 6**. The correlation coefficient of the isotherm exhibits a moderate level of association. It is possible to predict the effectiveness of the adsorption process using the dimensionless equilibrium parameter, R_L [53]. It is calculated as $R_L = 1/(1+b\, C_i)$, where b and C_i represent the adsorption constant and initial concentration (mg/L) of the solute, respectively. According to literature [54], the isotherm's shape can be categorized as unfavorable ($R_L > 1$), linear ($R_L = 1$), favorable ($0 < R_L < 1$), or irreversible ($R_L = 0$), based on the value of R_L. **Figure 9(b)** displays the R_L values that were computed for varying initial concentrations of Pb^{+2}. The range of R_L values falls within the interval of 0 to 1, thus indicating the favorable nature of the adsorption process [55]. Monolayer adsorption capacity was observed to be high (1951.98 mg/g) for CF2 compared to CF1 (1382.74 mg/g). The Langmuir isotherm plots were not perfectly linear as evidenced by the moderate R^2 values (see **Table 6**), which represent that Pb^{+2} ions adsorption is not completely homogenous monolayer surface adsorption process for both the magnetic adsorbents. Thus, isotherm data were further analyzed using Freundlich isotherm (see **Figure 9(c)**), which [50, 56] assumes that adsorption occurs on a heterogeneous surface through a multilayer adsorption

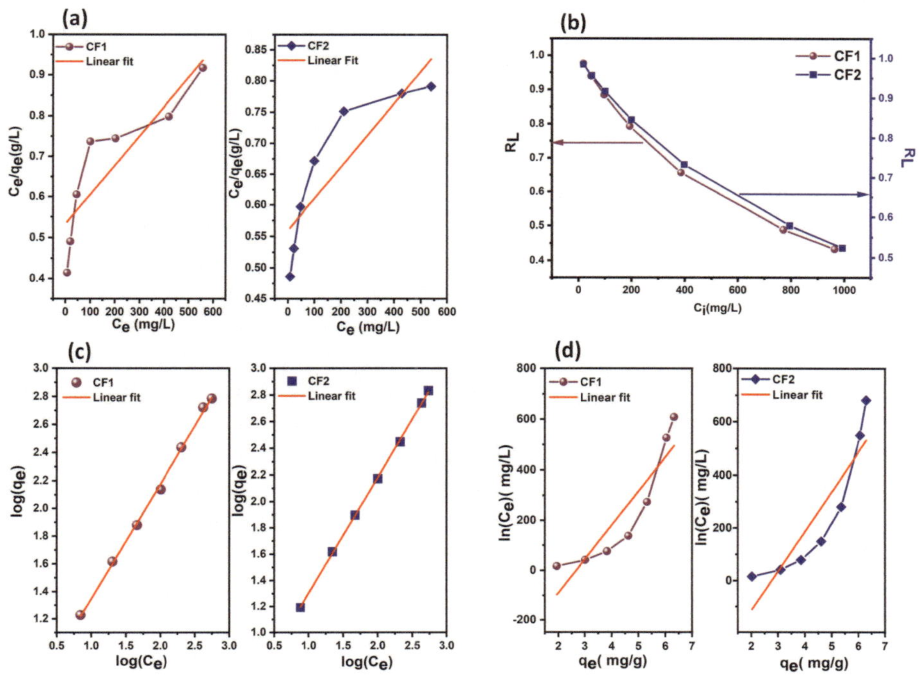

Figure 9.
Equilibrium Langmuir isotherms of Pb^{+2} ions by CF1 and CF2 adsorbents (a), calculated R_L values at different initial concentrations of Pb^{+2} for CF1 and CF2 (b), equilibrium Freundlich isotherms of Pb^{+2} ions by CF1 and CF2 (c), and Tempkin adsorption isotherm plots for CF1 and CF2.

Langmuir			
Sample	$b \times 10^{-3}$ **(g.mg^{-1})**	q_m **(mg.g^{-1})**	**R^2**
CF1	1.360	1382.74	0.727
CF2	0.917	1951.98	0.730
Freundlich			
Sample	k_f (mg.g^{-1})	N	R^2
CF1	3.30	1.21	0.998
CF2	2.67	1.14	0.999
Tempkin			
Sample	k_1 (L/g)	k_2	R^2
CF1	135.31	0.069	0.7893
CF2	150.31	0.063	0.7764

Table 6.
Characteristics of Langmuir, Freundlich, and Tempkin isotherm for the adsorption of Pb^{+2} onto the CF1 and CF2 along with the R^2 value.

mechanism and that the adsorbed amount increases with concentration according to the Freundlich isotherm. According to **Table 6**, it is observed that the R^2 values are very close to 1, which indicates that the experimental adsorption isotherms are very well modeled by the Freundlich equation compared to Langmuir model. Meanwhile, the value of n for Freundlich model is greater than 1, indicating the adsorption of Pb^{+2} also exhibits a favorable shape. Adsorption is considered satisfactory when the Freundlich constant n takes values within the range 1–10. Furthermore, Tempkin plots for magnetic adsorbents are shown in **Figure 9(d)** and Tempkin isotherm constant are listed in **Table 6**. The values of R^2 are greater than the data modeled by Langmuir isotherm but less than the Freundlich isotherm. Importantly, the heat of adsorption of microgranules is higher (150.31 L/g) than the nano-adsorbent (135.31 L/g).

From the above discussion, the overall adsorption of Pb^{+2} ions occurs through a multilayer adsorption mechanism for both the magnetic adsorbents. Moreover, the presence of adsorbate-adsorbate interactions was also observed during Pb^{2+} adsorption process, as checked by Tempkin isotherm. However, on the basis of R^2 value, the order of kinetic model followed for nano-adsorbent is as follows for the experimental data; Pseudo-second-order > Elovich model > Bangham's pore diffusion model > pseudo-first-order > intraparticle diffusion model. On the other hand, for microgranules adsorbent the order of kinetic model is as follows; intraparticle diffusion model> Bangham's pore diffusion model> pseudo-second-order> pseudo-first-order> Elovich model. Importantly, monolayer adsorption capacity observed to be high (1951.98 mg/g) for CF2 compared to CF1 (1382.74 mg/g) along with higher heat of adsorption for CF2 (150.31 L/g) than the CF1 (135.31 L/g), which suggest that the adsorption capacity of nanostructured CFO microgranules can be enhanced further by various modifications.

3.5.3 Regeneration study

In five consecutive cycles, the regeneration and re-adsorption of magnetic adsorbents showed a 99% regeneration rate, indicating that donor sites on the surface of

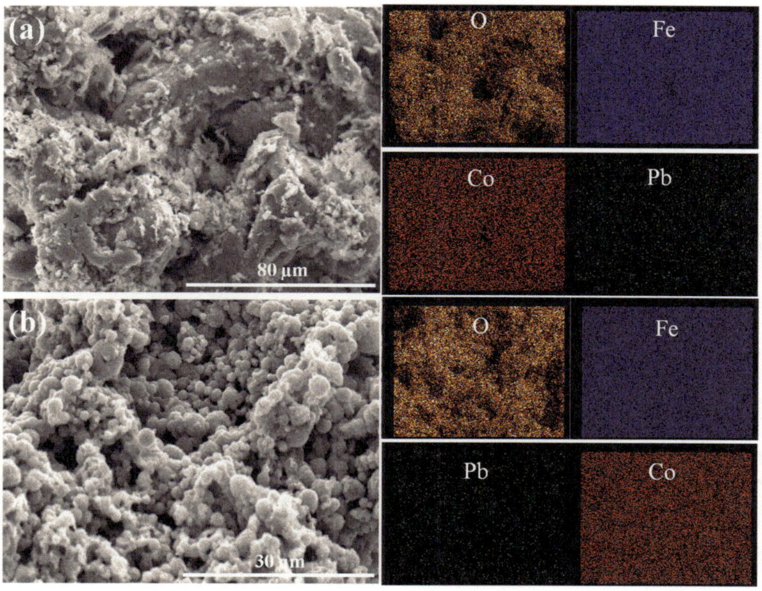

Figure 10.
Elemental distribution of Co, Fe, O, and Pb after the adsorption study for CF1 (a) and CF2 (b).

magnetic adsorbents and Pb^{+2} ions are reversible. In conclusion, magnetic absorbents and Pb^{+2} ions did not form strong bonds. Thus, the whole adsorption and desorption process does not include chemical redox reactions. Interestingly, the adsorption capacity of CF2 with greater size drops significantly during desorption with the rise in cycle number due to its low adsorption rate and small capacity.

Figure 10 display the SEM micrographs and elemental mappings of both adsorbents following adsorption and desorption experiments. The morphology of both adsorbents is preserved. The observation of a uniform distribution of Pb^{+2} ions adsorbed on the MNPs confirmed the adsorption of Pb^{+2} by the MNPs. Pb^{+2} was, however, preferentially adsorbed on the particle's surface rather than in its substance. As anticipated, the quantification of the elements confirmed that the CFO NPs were the source of the high concentrations of Co, Fe, and O. Evidently, the relatively low concentration of Pb^{+2} was produced by ion adsorption on the surface of MNPs. Significantly, CF1 contained higher concentrations of Pb + 2 ions than CF2.

4. Conclusions

We have successfully synthesized the novel CFO NPs and nanostructured CFO microgranules and systematically investigated their physicochemical properties. Our results show that both the CFO nanoparticles and nanostructured CFO microgranules favors inverse spinel structure with spherical and quasi-spherical morphology. For nanostructured CFO microgranules, M_S value is remarkably high (5.62 μ_B/F.U.) compared to the bulk counterpart, and almost double as compared to M_S value reported for CFO NPs. Our studies show that overall, the adsorption of Pb^{+2} ions occurs through a multilayer adsorption mechanism for magnetic nano- and nanostructured micro-adsorbents. Moreover, the existence of adsorbate-adsorbate interactions was also observed during Pb^{2+} adsorption process as checked by Tempkin isotherm.

Monolayer adsorption capacity was observed to be high (1951.98 mg/g) for nano-structured micro-adsorbents compared to nano-adsorbents (1382.74 mg/g) along with higher heat of adsorption of nanostructured micro-adsorbents (150.31 L/g) than the nano-adsorbents (135.31 L/g), which suggests that the adsorption capacity of nano-structured CFO microgranules can be enhanced further by various modifications. The proposed magnetic nano-adsorbents and nanostructured microgranules can be successfully applied for the removal of other heavy metal ions from aqueous solutions and complex industrial wastes.

Acknowledgements

S. M. Ansari gratefully acknowledges the financial support from BARC, Mumbai (Grant code: GOI–E-175). We thankfully appreciate the support of Dr. R. S. Devan, Metallurgy Engineering and Materials Science, Indian Institute of Technology (IIT), Indore, India in obtaining the FESEM micrographs. We thankfully appreciate the technical support of Mr. N. Patil and Dr. A. Supekar, Department of Geology, Savitribai Phule Pune University, Pune, India for providing the atomic absorption spectroscopy facility. The authors are also thankful to Prof. S. J. Sangode, Department of Geology, Savitribai Phule Pune University, Pune for providing BET measurements facility. The authors at the University of Texas at El Paso acknowledge, with pleasure, support from the National Science Foundation (NSF) with NSF-PREM grant #DMR-1827745.

Conflict of interest

The authors declare no conflict of interest.

Author details

Sumayya M. Ansari[1*], Vikas Kashid[2], Bhavesh B. Sinha[3], Debasis Sen[4,5], Yesh D. Kolekar[1] and Chintalapalle V. Ramana[6]

1 Department of Physics, Savitribai Phule Pune University, Pune, India

2 MIE-SPPU Institute of Higher Education, Doha-Qatar

3 Center for Nanoscience and Nanotechnology, University of Mumbai, Mumbai, India

4 Solid State Physics Division, Bhabha Atomic Research Centre, Mumbai, India

5 Homi Bhabha National Institute, Mumbai, India

6 Center for Advanced Materials Research (CMR), University of Texas at El Paso, El Paso, Texas, USA

*Address all correspondence to: smansari777@gmail.com

IntechOpen

© 2023 The Author(s). Licensee IntechOpen. This chapter is distributed under the terms of the Creative Commons Attribution License (http://creativecommons.org/licenses/by/3.0), which permits unrestricted use, distribution, and reproduction in any medium, provided the original work is properly cited.

References

[1] Ansari SM, Sinha BB, Pai KR, Bhat SK, Ma Y-R, Sen D, et al. Controlled surface/interface structure and spin enabled superior properties and biocompatibility of cobalt ferrite nanoparticles. Applied Surface Science. 2018;**459**:788-801. DOI: 10.1016/j.apsusc.2018.08.063

[2] Ansari SM, Sinha BB, Sen D, Sastry PU, Kolekar YD, Ramana CV. Effect of oleylamine on the surface chemistry, morphology, electronic structure, and magnetic properties of cobalt ferrite nanoparticles. Nanomaterials. 2022;**12**:3015

[3] Pan J, Hu P, Guo Y, Hao J, Ni D, Xu Y, et al. Combined magnetic hyperthermia and immune therapy for primary and metastatic tumor treatments. ACS Nano. 2020;**14**:1033-1044. DOI: 10.1021/acsnano.9b08550

[4] Zhang J, Zhu S, Xia W, Ming J, Li F, Fu J. Micromagnetic configuration of variable nanostructured cobalt ferrite: Modulating and simulations toward memory devices. ACS Applied Materials & Interfaces. 2019;**11**:28442-28448. DOI: 10.1021/acsami.9b07502

[5] Li J, Zhou Q, Zhong C, Li S, Shen Z, Pu J, et al. $(Co/Fe)_4O_4$ Cubane-containing nanorings fabricated by phosphorylating cobalt ferrite for highly efficient oxygen evolution reaction. ACS Catalysis. 2019;**9**:3878-3887. DOI: 10.1021/acscatal.9b00293

[6] Long L, Yang E, Qi X, Xie R, Bai Z-C, Qin S, et al. Positive and reverse core/shell structure $Co_xFe_{3-x}O_4/MoS_2$ and $MoS_2/Co_xFe_{3-x}O_4$ nanocomposites: Selective production and outstanding electromagnetic absorption comprehensive performance. ACS Sustainable Chemistry & Engineering. 2020;**8**:613-623. DOI: 10.1021/acssuschemeng.9b06205

[7] Pardo A, Yáñez S, Piñeiro Y, Iglesias-Rey R, Al-Modlej A, Barbosa S, et al. Cubic anisotropic Co- and Zn-substituted ferrite nanoparticles as multimodal magnetic agents. ACS Applied Materials & Interfaces. 2020;**12**: 9017-9031. DOI: 10.1021/acsami.9b20496

[8] Khanra S, Abdullah-Al Mamun M, Ferreira FF, Ghosh K, Guha S. Functionalized self-assembled peptide nanotubes with cobalt ferrite nanoparticles for applications in organic electronics. ACS Applied Nano Materials. 2018;**1**:1175-1187. DOI: 10.1021/acsanm.7b00344

[9] Ansari SM, Sinha BB, Phase D, Sen D, Sastry PU, Kolekar YD, et al. Particle size, morphology, and chemical composition controlled CoFe2O4 nanoparticles with tunable magnetic properties via oleic acid based solvothermal synthesis for application in electronic devices. ACS Applied Nano Materials. 2019;**2**:1828-1843. DOI: 10.1021/acsanm.8b02009

[10] Ansari SM, Suryawanshi SR, More MA, Sen D, Kolekar YD, Ramana CV. Field emission properties of nano-structured cobalt ferrite (CoFe2O4) synthesized by low-temperature chemical method. Chemical Physics Letters. 2018;**701**:151-156. DOI: 10.1016/j.cplett.2018.04.027

[11] Morales-Amaya CG, Alarcón-Herrera MT, Astudillo-Sánchez PD, Lozano-Morales SA, Licea-Jiménez L, Reynoso-Cuevas L. Ferrous magnetic nanoparticles for arsenic removal from groundwater. Water. 2021;**13**:2511

[12] Verma B, Balomajumder C. Fabrication of magnetic cobalt ferrite nanocomposites: An advanced method of removal of toxic dichromate ions from electroplating wastewater. Korean Journal of Chemical Engineering. 2020; **37**:1157-1165. DOI: 10.1007/s11814-020-0516-3

[13] Amiri M, Salavati-Niasari M, Akbari A, Gholami T. Removal of malachite green (a toxic dye) from water by cobalt ferrite silica magnetic nanocomposite: Herbal and green sol-gel autocombustion synthesis. International Journal of Hydrogen Energy. 2017;**42**: 24846-24860. DOI: 10.1016/j.ijhydene.2017.08.077

[14] Jayalakshmi R, Jeyanthi J, Aswin Sidhaarth KR. Versatile application of cobalt ferrite nanoparticles for the removal of heavy metals and dyes from aqueous solution environmental nanotechnology. Monitoring & Management. 2022;**17**:100659. DOI: 10.1016/j.enmm.2022.100659

[15] Santhosh C, Kollu P, Felix S, Velmurugan V, Jeong SK, Grace AN. $CoFe_2O_4$ and $NiFe_2O_4$@graphene adsorbents for heavy metal ions – Kinetic and thermodynamic analysis. RSC Advances. 2015;**5**:28965-28972. DOI: 10.1039/C5RA02905H

[16] Al-Faiyz YSS, Gouda M. Multi-Walled carbon nanotubes functionalized with hydroxamic Acid derivatives for the removal of lead from wastewater: kinetics, isotherm, and thermodynamic studies. Polymers. 2022;**14**:3870

[17] Shao D, Chen C, Wang X. Application of polyaniline and multiwalled carbon nanotube magnetic composites for removal of Pb(II). Chemical Engineering Journal. 2012; **185-186**:144-150. DOI: 10.1016/j.cej.2012.01.063

[18] Singh S, Barick KC, Bahadur D. Surface engineered magnetic nanoparticles for removal of toxic metal ions and bacterial pathogens. Journal of Hazardous Materials. 2011;**192**: 1539-1547. DOI: 10.1016/j.jhazmat.2011.06.074

[19] Zhang D, Wei S, Kaila C, Su X, Wu J, Karki AB, et al. Carbon-stabilized iron nanoparticles for environmental remediation. Nanoscale. 2010;**2**:917-919. DOI: 10.1039/C0NR00065E

[20] Yang X, Lee J, Yuan L, Chae S-R, Peterson VK, Minett AI, et al. Removal of natural organic matter in water using functionalised carbon nanotube buckypaper. Carbon. 2013;**59**:160-166. DOI: 10.1016/j.carbon.2013.03.005

[21] Zhou L, Ji L, Ma PC, Shao Y, Zhang H, Gao W, et al. Development of carbon nanotubes/$CoFe_2O_4$ magnetic hybrid material for removal of tetrabromobisphenol a and Pb(II). Journal of Hazardous Materials. 2014; **265**:104-114. DOI: 10.1016/j.jhazmat.2013.11.058

[22] Chen C, Hu J, Shao D, Li J, Wang X. Adsorption behavior of multiwall carbon nanotube/iron oxide magnetic composites for Ni(II) and Sr(II). Journal of Hazardous Materials. 2009;**164**: 923-928. DOI: 10.1016/j.jhazmat.2008.08.089

[23] Viltužnik B, Košak A, Zub YL, Lobnik A. Removal of Pb(II) ions from aqueous systems using thiol-functionalized cobalt-ferrite magnetic nanoparticles. Journal of Sol-Gel Science and Technology. 2013;**68**:365-373. DOI: 10.1007/s10971-013-3072-z

[24] Li K, Xie L, Hao Z, Xiao M. Effective removal of Hg(II) ion from aqueous solutions by thiol functionalized cobalt ferrite magnetic mesoporous silica

composite. Journal of Dispersion Science and Technology. 2020;**41**:503-509. DOI: 10.1080/01932691.2019.1591974

[25] Qu S, Huang F, Yu S, Chen G, Kong J. Magnetic removal of dyes from aqueous solution using multi-walled carbon nanotubes filled with Fe_2O_3 particles. Journal of Hazardous Materials. 2008;**160**:643-647. DOI: 10.1016/j.jhazmat.2008.03.037

[26] Gong JL, Wang B, Zeng GM, Yang CP, Niu CG, Niu QY, et al. Removal of cationic dyes from aqueous solution using magnetic multi-wall carbon nanotube nanocomposite as adsorbent. Journal of Hazardous Materials. 2009;**164**:1517-1522. DOI: 10.1016/j.jhazmat.2008.09.072

[27] Yang N, Zhu S, Zhang D, Xu S. Synthesis and properties of magnetic Fe_3O_4-activated Carbon nanocomposite particles for dye removal materials letters. Materials Letters. 2008;**62**: 645-647. DOI: 10.1016/j.matlet.2007.06.049

[28] Rocher V, Siaugue JM, Cabuil V, Bee A. Removal of organic dyes by magnetic alginate beads. Water Research. 2008;**42**:1290-1298. DOI: 10.1016/j.watres.2007.09.024

[29] Ansari SM, Kashid V, Salunke H, Sen D, Kolekar YD, Ramana C. First-principles calculations of the electronic structure and magnetism of nanostructured Co Fe 2 O 4 microgranules and nanoparticles. Physical Review B. 2020;**102**:035446

[30] Li N, Zheng M, Chang X, Ji G, Lu H, Xue L, et al. Preparation of magnetic $CoFe_2O_4$-functionalized graphene sheets via a facile hydrothermal method and their adsorption properties. Journal of Solid State Chemistry. 2011;**184**:953-958. DOI: 10.1016/j.jssc.2011.01.014

[31] Ai L, Huang H, Chen Z, Wei X, Jiang J. Activated Carbon/$CoFe_2O_4$ composites: Facile synthesis, magnetic performance and their potential application for the removal of malachite green from water chemical. Engineering Journal - Chemical Engineering Journal. 2010;**156**:243-249. DOI: 10.1016/j.cej.2009.08.028

[32] Farghali AA, Bahgat M, El Rouby WMA, Khedr MH. Decoration of MWCNTs with $CoFe_2O_4$ nanoparticles for methylene blue dye. Adsorption Journal of Solution Chemistry. 2012;**41**: 2209-2225. DOI: 10.1007/s10953-012-9934-0

[33] Deng Y, Qi D, Deng C, Zhang X, Zhao D. Superparamagnetic high-magnetization microspheres with an Fe_3O_4@SiO_2 core and perpendicularly aligned mesoporous SiO_2 shell for removal of microcystins. Journal of the American Chemical Society. 2008;**130**: 28-29. DOI: 10.1021/ja0777584

[34] Ansari SM, Sen D, Haritha K, Kolekar YD, Low-cost RCV. Ecofriendly, and large-scale synthesis of nanostructured $Co_{1-x}Mn_xFe_2O_4$ microgranules with enhanced magnetic performance by chemical spray drying processing. Colloids and Surfaces A: Physicochemical and Engineering Aspects. 2023;**672**:131697. DOI: 10.1016/j.colsurfa.2023.131697

[35] Ansari SM, Bhor RD, Pai KR, Mazumder S, Sen D, Kolekar YD, et al. Size and chemistry controlled cobalt-ferrite nanoparticles and their anti-proliferative effect against the MCF-7 breast cancer cells ACS. Biomaterials Science & Engineering. 2016;**2**: 2139-2152. DOI: 10.1021/acsbiomaterials.6b00333

[36] Guinier A, Fournet G, Walker CB, Vineyard GH. Small-angle scattering of

X-rays. Physics Today. 1956;**9**:38-39. DOI: 10.1063/1.3060069

[37] Pedersen JS. Determination of size distribution from small-angle scattering data for systems with effective hard-sphere interactions. Journal of Applied Crystallography. 1994;**27**:595-608. DOI: 10.1107/S0021889893013810

[38] Monárrez-Cordero BE, Amézaga-Madrid P, Leyva-Porras CC, Pizá-Ruiz P, Miki-Yoshida M. Study of the Adsorption of Arsenic (III and V) by Magnetite Nanoparticles Synthetized via AACVD. Materials Research. 2016;**19**: 103-112

[39] Glatter O, Kratky O, editors. Small Angle x-ray Scattering. London; New York: Academic Press; 1982

[40] Telser LG. The lognormal distribution, J. Aitchison and J. A. C. Brown, Cambridge, England: Cambridge University Press, 1957, pp. xviii, 176. $6.50. American Journal of Agricultural Economics. 1959;**41**:161-162. DOI: 10.2307/1235218

[41] Teixeira J. Small-angle scattering by fractal systems. Journal of Applied Crystallography. 1988;**21**:781-785. DOI: 10.1107/S0021889888000263

[42] Baxter RJ. Percus–Yevick equation for hard spheres with surface adhesion. The Journal of Chemical Physics. 1968; **49**:2770-2774. DOI: 10.1063/1.1670482

[43] Sing KSW. Reporting physisorption data for gas/solid systems with special reference to the determination of surface area and porosity (recommendations 1984). Pure and Applied Chemistry. 1985;**57**:603-619. DOI: 10.1351/pac198557040603

[44] Shirsath DS, Shirivastava VS. Adsorptive removal of heavy metals by magnetic nanoadsorbent: An equilibrium and thermodynamic study applied. Nano. 2015;**5**:927-935. DOI: 10.1007/s13204-014-0390-6

[45] Iqbal J, Wattoo FH, Wattoo MHS, Malik R, Tirmizi SA. Imran M and Ghangro a B Adsorption of acid yellow dye on flakes of chitosan prepared from fishery wastes. Arabian Journal of Chemistry. 2011;**4**:389-395. DOI: 10.1016/j.arabjc.2010.07.007

[46] Kuyucak N, Volesky B. Accumulation of cobalt by marine alga. Biotechnology and Bioengineering. 1989; **33**:809-814. DOI: 10.1002/bit.260330703

[47] Borhade AV, Kankrej S. R an efficient cost-effective removal of Ca^{2+}, Mg^{2+}, and Cu^{2+} ions from aqueous medium using chlorosodalite synthesized from coal fly ash. Journal of Chemical & Engineering Data. 2017;**62**: 596-607. DOI: 10.1021/acs.jced.6b00588

[48] Bayazit ŞS, Kerkez Ö. Hexavalent chromium adsorption on superparamagnetic multi-wall carbon nanotubes and activated carbon composites. Chemical Engineering Research & Design. 2014;**92**:2725-2733

[49] Boyd GE, Adamson AW, Myers LS, Jr. The exchange Adsorption of ions from aqueous solutions by organic zeolites. II. Kinetics1. Journal of the American Chemical Society. 1947;**69**: 2836-2848. DOI: 10.1021/ja01203a066

[50] Ai L, Li M, Li L. Adsorption of methylene blue from aqueous solution with activated carbon/cobalt ferrite/alginate composite beads: Kinetics, isotherms, and thermodynamics. Journal of Chemical & Engineering Data. 2011; **56**:3475-3483. DOI: 10.1021/je200536h

[51] Chien SH, Clayton WR. Application of Elovich equation to the kinetics of

phosphate release and sorption in soils. Soil Science Society of America Journal. 1980;**44**:265-268. DOI: 10.2136/sssaj1980.03615995004400020013x

[52] Langmuir I. The adsorption of gases on plane surfaces of glass, mica and platinum. Journal of the American Chemical Society. 1918;**40**:1361-1403. DOI: 10.1021/ja02242a004

[53] Weber TW, Chakravorti RK. Pore and solid diffusion models for fixed-bed adsorbers. AICHE Journal. 1974;**20**: 228-238. DOI: 10.1002/aic.690200204

[54] McKay G. Adsorption of dyestuffs from aqueous solutions with activated carbon I: Equilibrium and batch contact-time studies. Journal of Chemical Technology and Biotechnology. 1982;**32**: 759-772. DOI: 10.1002/jctb.5030320712

[55] Xiong L, Yang Y, Mai J, Sun W, Zhang C, Wei D, et al. Adsorption behavior of methylene blue onto titanate nanotubes. The Chemical Engineering Journal. 2009;**156**:313. DOI: 10.1016/j.cej.2009.10.023

[56] Langmuir I. The constitution and fundamental properties of solids and liquids. Part I. Solid. Journal of the American Chemical Society. 1916;**38**: 2221-2295. DOI: 10.1021/ja02268a002

Chapter 2

Applications of Nano-Ferrites in Medicine

Amina Ibrahim Ghoneim

Abstract

Nano-ferrites are elegant and smart nanoparticles. Biomedical implementations of nano-ferrites include cell signaling, hyperthermia, magnetic resonance imaging (MRI), nanorobots, drug delivery, anti-cancer function, anti-fungi, anti-bacteria, biosensors, brain stimulation, wound healing, etc. Nano-ferrites showed anticancer features towards various types of cancer cells, such as breast cancer. They have been used for drug delivery as well as drug release. Magnetic hyperthermia is a successful approach in cancer treatment, where nano-ferrites have been used under the influence of external magnetic fields. Nano-ferrites are used as magnetic resonance imaging (MRI) contrast agents. Furthermore, nano-ferrites have been involved in the magnetic nanorobots and biosensors industries. Superparamagnetic nanocrystals including manganese nano-ferrites have been utilized in the activation of thermos-sensitive transient receptor potential vanilloid 1 (TRPV1) channels to attain further brain stimulation. The chemo-genetic modulation of targeted neural circuits using superparamagnetic nano-ferrite particles provides a novel strategy for brain stimulation so as to investigate brain functions and neurological diseases. Moreover, they have antibacterial and antifungal activity against several types of bacteria and fungi, such as *E. coli* and *Candida albicans*. Pr_6O_{11}/Fe_2O_3/GO@PCL film nanocomposites have been used in skin wound healing treatment; thence, these smart new nanocomposites could be utilized in the advancement of wound healing applications.

Keywords: biomedical applications, cell signaling, hyperthermia, magnetic resonance imaging (MRI), nanorobots, drug delivery, anti-cancer function, anti-fungi, anti-bacteria, biosensors, brain stimulation, wound healing

1. Introduction

Nano-ferrites with their wide variety of types continue to be extremely essential nanocrystals in the recent era from scientific, industrial as well as technological prospects ever since their discovery in 1950s. Specification of such nanocrystals as nano-spinel, nano-hexaferrites, and orthoferrites stems from their tiny nano-particles size as well as their Versatile crystal structures. Recently, ultimate advanced scientific research vistas have essentially focused on promoting and preparing new unique nanocrystals with ultra-fine grain sizes for utilizations in too diverse and spreading technological fields likespintronics, magnetic recording media, shielding technology, and microwave attenuators imputing to their extremely high thermal stability, non-toxic merits, supreme chemical stability and extremely supreme corrosion resistivity [1–5]. Obviously, Ferrite nanocrystals are vital in industrial vistas such as their uses in wireless networks, Mobile technologies, RADAR, and permanent magnets. The ultimate focus of researchers today is synthesizing these nanocrystals utilizing diverse synthesis procedures and varied characterizing strategies [5–8].

Early in history, ancient Greeks used lodestone (magnetite Fe_3O_4) to cure many diseases. By precisely advanced nano-biotechnology, utilization of these diverse magnetic ferrite nanocrystals has emerged, which extends to clinical research vistas and modern biomedicine aspects. Since biological tissues are transparent to magnetic fields, in biomedical implementations ferrite nanocrystals could perform various functions synchronized with invisible tissue penetrating magnetic fields. For example, a sort of natural magnetotactic bacteria naturally form Fe oxide nanocrystals in order to generate permanent diodes interior their own natural cells so as to navigate towards the most convenient habitats by the outstanding property of sensing our planet's magnetic field. These unique ferrite nanocrystals obtain excellent biocompatible features, which in turn used as therapeutic and magnetic resonance imaging agents in clinical scientific research. Then, for the supreme magneto-responsive merits of these nanocrystals, versatile expanding biomedical implementations are created. These wide expanding uses include magnetic resonance imaging (MRI) for various tissues, nanorobots, cell signaling, curing of illnesses via hyperthermia, drug delivery, anti-cancer function, anti-bacterial and anti-fungi function, biosensors, etc. Obviously, ferrite nanocrystals can be delivered to the targeted biological tissues, thus they possess the capability to induce diverse cell responses for modulating biological systems via conversion of external magnetic field energies into bio-sensitive signals, such as heating stimuli, mechanical forces and inducing local field [9–15]. However, nanocrystal function is highly influenced by the magnetic features of these ferrite ultrafine nanoparticle mediators. Therefore, for efficient transduction of biochemical signals, it is extremely vital to engineer ferrite nanocrystals with enhanced magnetic response. In this respect, biological effects generated by ferrite nanocrystals are extremely influenced by their magnetic merits like saturation magnetization, chemical stability, thermal stability, corrosion resistivity, ultrafine size nature, chemical composition, crystalline structure, and surface morphology, which is closely related to thermal conversion efficiency. Furthermore, several strategies could be taken into consideration in synthesizing assemblies or clusters of magnetic ferrite nanoparticles for improving their performance and merit like saturation magnetization and enhancing their magnetic response for their pioneering biomedical implementations [9–15].

One of the most recent talented nanostructures is the one-dimensional magnetic nanowire, nanotube, and nanofiber, which possess unique chemical and physical merits. Thus, these magnetic nanofibers have several implementations such as nanoscale magnetic devices industry and biomedical applications like hyperthermia which are extensively attractive and low-cost strategies for cancer treatment. However, magnetic nanocrystals are utilized as supporting material in Radiotherapy and Chemotherapy. Iron oxide nanocrystalline diverse structures have been extensively utilized for hyperthermia application imputing to their ultimate non-toxicity, supreme corrosion resistivity, superior chemical stability, low cost, and echo-friendly habit. As an example, Ni ferrite nanofiber which can be synthesized via electrospinning procedure and also can be annealed to variable annealing temperature degrees in order to obtain various sizes of nanofibers, has been utilized for hyperthermia applications. Their shape enables them to produce extra heat energy at RF alternating magnetic fields. Oleic acid-coated Ni ferrite nanocrystals were utilized for hyperthermia and drug delivery applications. The scarcity of research on the synthesis, structural, magnetic, and hyperthermia of nickel ferrite nanofibers makes them marvelous nanostructures for these studies [16].

Nano-ferrites have diverse nanocrystalline structures and morphological shapes such as one-dimensional (1D) nanoparticles, two-dimensional (2D) nanoparticles, and three-dimensional (3D) nanoparticles. One-dimensional nanoparticles consist of nanorods, nanowires, and nanotubes, and possess obvious interesting biomedical implementations imputing to their unique magnetic merits. Nanorods possess diameters beginning from a few nanometers up to 100 nm in length. Nanotubes have hollow nanorods morphological shapes. Nanowires possess lengths greater than 100 nm. They can be synthesized via diverse preparation routes like hydrothermal, co-precipitation strategies, etc. While two-dimensional (2D) nanoparticles consist of nanofilms, nanoplates, and nanosheets. These 2-dimensional nanostructures could be synthesized via thermal decomposition and co-precipitation strategies with the precise control of annealing temperatures and conditions, furthermore, they are convenient candidates for several biomedical implementations. On the other hand, the anisotropic three-dimensional (3D) nanoparticles consist of nanospheres, nano-cubes, and nanoflowers. 3-Dimentional nanostructures could be synthesized via thermal decomposition and co-precipitation strategies, etc., with the accurate handling and control of annealing temperatures and conditions, over and above, they are convenient nanocrystals for several biomedical implementations [17]. These diverse morphological nanostructures of nano-ferrites are indicated in **Figure 1** [17].

Figure 1.
TEM images of different shapes of $MnFe_2O_4$ nanoparticles (a) nanospheres, (b) nano-cubes, (c) nanorods, (d) nanowires, (e) nanorods, (f) triangular, (g) polyhedron, and (h) nano-octahedron [17].

2. Biomedical applications of nano-ferrites

Biomedical implementations of nano-ferrites extensively dependent on the toxicity and biocompatibility of these nanocrystals. There are some considerable parameters that influence the toxicity and bio-compatibility of these ferrite nanoparticles such as morphological shape and size of nanoparticles including their coatings as well as their magnetic response. The main outstanding merits of nano-ferrite crystals in order to be utilized in various biomedical implementations in in vivo and in vitro are biodegradability, low toxicity, supreme magneto-responsive attribute, and long blood retention time. Obviously, nano-ferrites possess versatile expanding biomedical implementations including, cell signaling, hyperthermia, magnetic resonance imaging (MRI), nanorobots, drug delivery, anti-cancer function, anti-fungi, anti-bacteria, biosensors, etc. [11–17]. **Figure 2** illuminates ferrite nanoparticles and various polymer stabilization methods over them [17–19].

2.1 Toxicity impact

Toxicity impact in cell culture/blood components is a significant parameter that must be examined before any use of nano-ferrites in an in-vivo investigation. There exist several sorts of cytotoxicity examination such as Alamar Blue, MTT, and Trypan Blue assay, which lack standard procedure, hence several examination tests should be done. MTT colorimetric assay (3-[4,5-dimethylthiazol- 2-yl]-2,5 diphenyl tetrazolium bromide) is the most cost-effective and classical assay for cytotoxicity investigation. For the Alamar Blue assay, viable cells are found after whole number of cells are brood

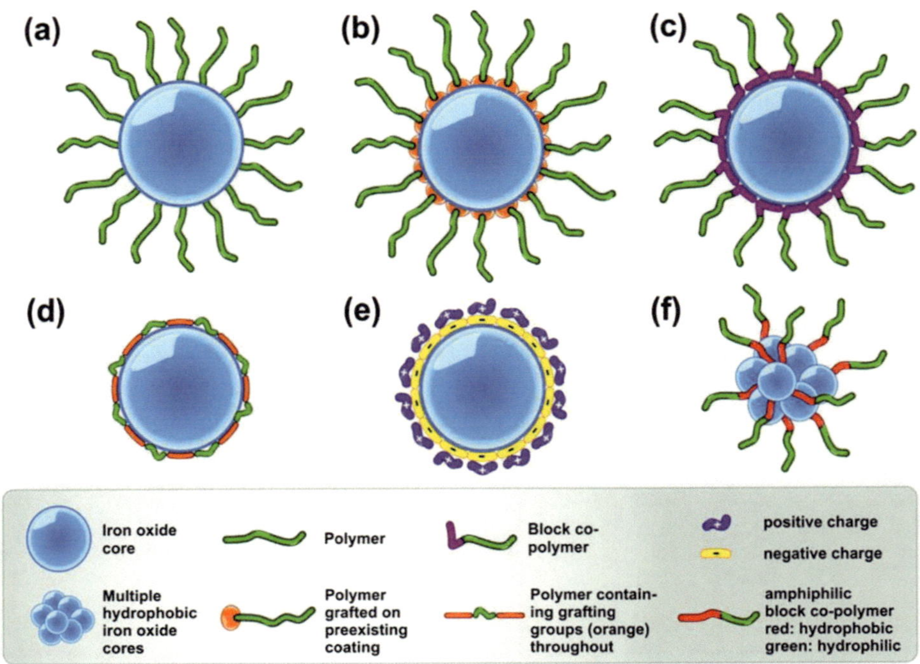

Figure 2.
Ferrite nanoparticles and various polymer stabilization methods over them [17].

with non-fluorescent resazurin dye. Exclusion test (Trypan blue dye assay) is used to determine dead cells. Reactive oxygen species (ROS) assay is utilized to examine the cytotoxicity of nano-ferrites. It is intended to determine ROS generation as an indicator of cellular oxidative stress. Nano-ferrites such as Mn ferrite nanoparticles are efficient nanocrystals for magnetic hyperthermia and their toxicity level was examined for various cancer cell lines like human melanoma cell line (MW35), mouse melanoma cell line (B16F10), human lung adenocarcinoma (A549) cell line, HeLa cell line, etc. As nano-ferrite concentration increases, cancer cell viability decreases [17, 18, 20]. Toxicity effect of $CoFe_2O_4$ nanoparticles (CF-NPs) and $Mg_{0.05}Co_{0.95}Fe_2O_4$ nanoparticles (MCF-NPs) on human breast cancer cells (MCF-7) via MTT assay indicated that, with the increase of ferrite nanocrystals concentration, the cytotoxicity was influenced and is dose-dependent, and the viability of the MCF-7 cells decreased as indicated in **Figure 3** [20].

The cytotoxic properties against human breast (MCF-7) cancer cells by MTT assay and their reactive oxygen species (ROS) are obviously showing an increased trend with the increase of concentration. It was obvious thatthe viability cells were dose-dependent. Cytotoxicity results occurred by decreasing cell viability with increasing nano-ferrites concentrations by the exposure to different kinds of Ferrite nanoparticles as indicated in **Figure 3**. The explanation for this is related to the entrance of NPs towards cancer cells, geometry, size, and distinctive properties of NPs were responsible for the cell death. Cell membranes exhibit small pore-like passages that facilitate to entry of ferrite nanoparticles into cells and damage to the upper layer of the cell membrane. The nanostructures can be easily entered into the cytoplasm of the cell and react with cell's organelles diminishing the growth of cancer cells. Further, morphological changes in cancer cells in the presence of Ferrite nanoparticles are occurring. Additionally, to know the cytotoxicity of cancer cells and the role of nano-ferrites in the mechanism, the reactive oxygen species (ROS) for 24 hours is also measured, as indicated in **Figure 4**. It was considered that the production ROS by interaction of nano-ferrites play the key role in mechanism of cytotoxicity because they can directly disturb the many components of cells such as DNA structure, proteins and lipids which in turn prime cause of cell death. Consequently, dose-dependent features and upsurge rate of ROS generation support the cytotoxicity results and cancer cell death as well [20].

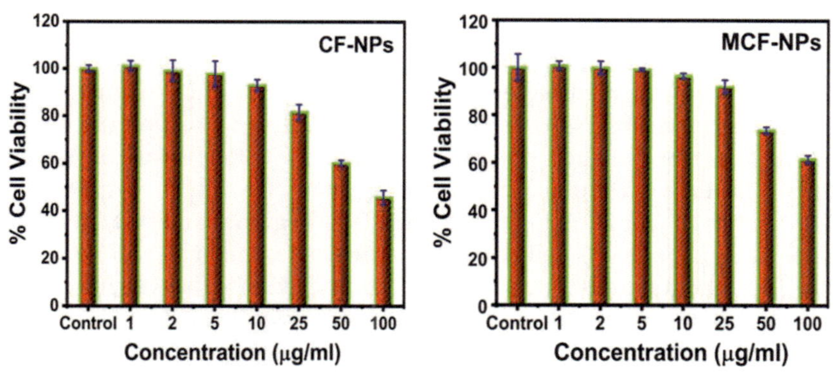

Figure 3.
Cytotoxicity (cancer cell death) of $CoFe_2O_4$ nanoparticles (CF-NPs) and $Mg_{0.05}Co_{0.95}Fe_2O_4$ nanoparticles (MCF-NPs) against human breast cancer cells MCF-7 [20].

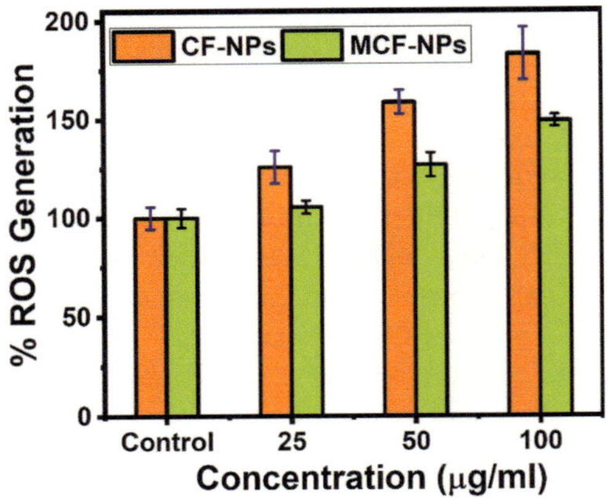

Figure 4.
The reactive oxygen species (ROS) generation by nano-ferrites for $CoFe_2O_4$ nanoparticles (CF-NPs) and $Mg_{0.05}Co_{0.95}Fe_2O_4$ nanoparticles (MCF-NPs) against in human breast cancer cells MCF-7 [20].

2.2 Drug delivery and release

Biocompatible nano-ferrites are distinguishable imputing to their drug delivery and release, tumor treatment capability, and safe liberation from bio-systems. Utilization of these nanocrystals reduces the required drug amount and avoids side effects, where nano-ferrite particles serve as a core covered with a shell of biocompatible organic moieties such as Chitosan, dextran, polyethylene-glycol, etc. Furthermore, a gold shell is utilized to protect nano-ferrites core from oxidation. Anticancer drug is embedded inside nanocomposites or bonded to the outer surface of nanocomposites via a linker. Nano-ferrite cores assist in guiding anticancer drugs via the use of external magnetic fields, whilst shells serve as surface modifiers or stabilizers at which they surge targeting potential. Drug release mechanisms are widely utilized, e.g., light, redox stimuli, enzyme, PH, thermal, etc. [20].

2.2.1 Drug delivery

Drug delivery agents like ferrite nanocrystals guided with external magnetic fields are the main focus for scientists from all over the world imputing to their efficient strategy, simple mechanism, simple synthesis procedure, cost-effectiveness, and targeting their accurate function in biological systems. These smart nanoparticles can be directed toward a specific cell or tissue maximizing their effect. Furthermore, nano-ferrites can deliver drugs like Dox and Cisplatin by encapsulating them inside a polymer matrix to the tumor tissue. They can also carry drugs circulate without leakage and move more flexibly to targeted tumors with aid of external magnetic field, thus assisting in giving an efficient therapeutic cure to the cancer. Once it is delivered to tumor tissue, the drug will be released and provide its therapeutic effects. Hybrids composed of nano-ferrites, anticancer drugs, semiconductors, and biocompatible coating agents (such as chitosan nano-ferrites hybrid with folate-conjugated tetrapeptide nanocomposite) are special agents for efficient cancer therapy [9, 20–22].

2.2.2 Drug release

The marvelously efficient drug release mechanisms, e.g., light, redox stimuli, enzyme, PH, thermal, etc., are extensively used. Another efficient procedure that also impacts drug release is using magnetic nano-ferrites with the aid of external magnetic fields. Magnetic guidance of drug release is obviously a marvelous route or strategy for controlled drug release. Then, nano-ferrites are extremely important agents acting in drug delivery and on the other hand, greatly impact drug release. Moreover, they are utilized in remote-controlled drug release. An example, for DOX loaded into nano-ferrite-chitosan as they are delivered to the target (tumor tissue) and influenced by alternative magnetic field, by heat generated by ferrite nanocrystals as well as the mechanical deformation occurring, DOX is easily released. As magnetic field amplitude surges the rate of DOX release increases up to an optimum rate of release. Furthermore, hyperthermia is a precise efficient strategy for drug delivery and release [20–22].

2.3 Magnetic hyperthermia

Cancer treatment strategies involve radiotherapy, chemotherapy, hyperthermia (HPT), and surgery. Magnetic HPT is an elegant procedure of cancer therapy that is very destructive for cancer cells with the aid of an external AC magnetic field using magnetic nanocrystals like nano-ferrites. Nanocrystals ratio and their toxic effects are the major anxieties in magnetic HPT. HPT focuses on the accumulation of guided magnetic nanoparticles at tumor site, and generation of heat under influence of AC magnetic fields, which in turn selectively kill tumor cells. Core-shell nanostructures are more efficient for magnetic HPT than single-phase nano-ferrites, at which self-heating merits under AC magnetic fields have been early examined for HPT. Types of HPT are localized, regional, and whole-body HPT. HPT is heating specific tumors up to 46°C, at which the normal enzyme processes are destroyed and blood vessels inside the tumor have low thermic resistance. HPT is an essential therapy for cancers, where cancers accept much more heat imputing to their surging metabolism rate. Control of exposure spans and heating rate is the only way to make HPT more precise and efficient via utilizing magnetic nanoparticles with a certain specific absorption rate (SAR) at low concentrations. SAR is the rate of heat generated per unit mass, which is highly influenced by the magnetic nanoparticles size, distribution, shape, anisotropy constant, saturation magnetization, and morphology. Generated heat depends on nanoparticle size and type as well as the frequency and amplitude of the AC magnetic field [17, 20, 23]. The conducting strategies for thermotherapy are utilizing microwave, laser therapy, and ionizing radiation. Nanoparticles are commonly utilized for thermotherapy for elimination of cancer cells, such as gold nanotubes, nano-ferrites, etc. Gold nanotubes convert IR rays into thermal energy for the treatment of tumors. Cobalt nano-ferrite possesses higher anisotropy and can lose magnetic moment slowly, thus, tumor cells could absorb the resultant heat more robustly. However, graphene nanosheets conjugated Cobalt nano-ferrites have lower cytotoxicity. Furthermore, graphene platelets prevent aggregation of Cobalt nano-ferrites particles without reduction of their magnetic potency. Graphene/cobalt ferrite nanoparticles are utilized for magnetic HPT implementations. Biocompatibility of graphene oxide (GO)/cobalt ferrite nanoparticles on neural stem cells has been investigated indicating that these nanostructures are excellent candidates for HPT [24].

2.4 Magnetic resonance imaging (MRI)

Magnetic resonance imaging (MRI) is a delicate diagnostic strategy that can provide 3-dimensional anatomical images of the body. MRI of tumors is the first stage in cancer therapy for locating tumor position, its spread extent, and determining curing strategy. MRI is used for revealing the early stages of various tumors and tracking drug responses. The principle of nuclear magnetic resonance (NMR) is the fundamental basis for MRI, which produces extremely high-accuracy biological images. Magnetic nano-ferrite particles have been extensively utilized as MRI contrast agents. MRI signal promotion is imputed to super-exchange interactions inside nano-ferrites as well as dipole-dipole interaction between nano-ferrites and H_2O surrounding them, to limit longitudinal or transverse relaxation times. Superparamagnetic nano-ferrites have been extensively utilized for in vivo biomedical implementations, like drug delivery, hyperthermia, and MRI contrast agents. Cobalt nano-ferrites proved excellent efficiency in MRI imputing their rise in anisotropy constant, coercivity, mild saturation magnetization, extremely high chemical stability, and biocompatible merits (as illuminated in **Figure 5**) [9, 25].

Iron oxides like Fe_3O_4 and Fe_2O_4 have been widely used as contrast agents due to their chemical stability, no toxicity, and biodegradability. Co-ferrite are considered good candidates for developing T_2 contrast agents with higher relaxivity. Basically, Co-ferrite ($CoFe_2O_4$) nanoparticles perturbed the magnetic relaxation process of protons in the tissue and induced the shortening of the spin–spin relaxation time of the proton. The high saturation magnetization leads to the relaxivity enhancement of undergone tissue of the body. MRI study of Co-ferrites was performed at low field MRI unit (0.35 Tesla MAGNETOM) in order to get high contrast underwent organs such as the liver and spleen of rabbits. For this purpose, a solution 0.02 mg Fe/Kg of iron oxides and cobalt ferrites was prepared in saline water, and an intravenous dose of contrast agent was delivered through vein of ear. Four rabbits of equal weight were taken to compare the contrast enhancement on MRI images. Using IQ View software, it has been measured that the intensity of signal induced by contrast agents such as

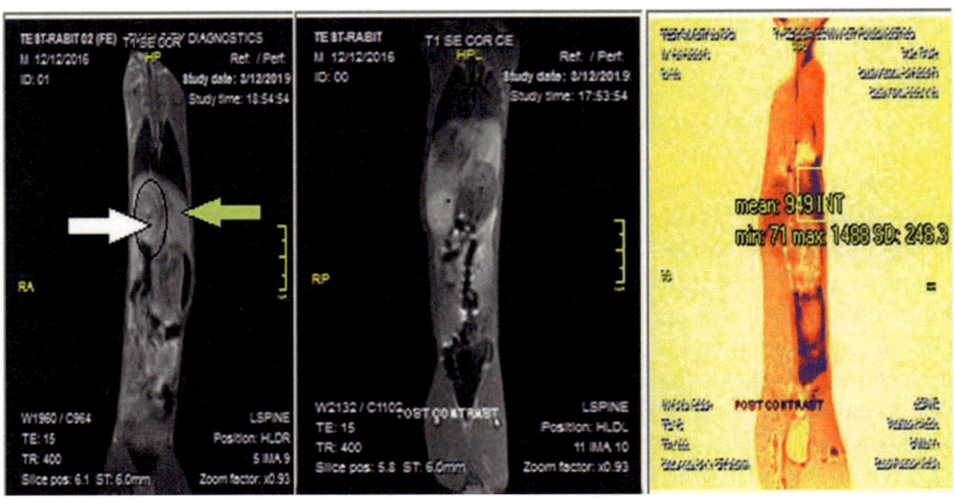

Figure 5.
MRI images of liver and spleen (a) CA Fe_3O_4, (b) CA cobalt ferrites—20%wt, (c) CA cobalt ferrites—40%wt. (white arrow show anatomy of stomach of rabbit, green arrow shows liver of rabbit) [25].

iron oxides (for liver, I = 1433 ± S.D. = 111.5, spleen, I = 1009 ± S.D. = 96.5), cobalt ferrites (for liver, I = 1513 ± S.D. = 102.2, spleen I = 1694 ± S.D. = 219.2) as shown in **Figure 5a–c**. It could be concluded that cobalt ferrites induced high signal intensity of T_2 contrast agents as compared to iron oxide nanoparticles. Metal-doped contrast agents may be used for clinical purposes to diagnose many diseases for clinical use of MR contrast agents [25].

2.5 Magnetic nanorobots

Magnetic nano-ferrites continue to surprise scientists in wide-spreading vistas of science, such as biomedicine, nano biosensors, magnetic nano-robots, etc. It could be collected as nanorobots of various styles, and magnetically guided to perform diagnosis and curing of illnesses. An example, a nanorobot has been created in a shape like a fan nanorobot covered by nano-liquid layer in order to precisely execute functions in the vitreous body of the eye and controlled in a delicate way. Also, a rectangular-shaped sheet magnetic robot has been fabricated from the hard magnetic NdFeB particles in silicon elastomer. Under the guidance of an external magnetic field over time, magnetic nanorobots could vary their appearance, climb, roll, walk, jump, crawl, and swim. Furthermore, they could grab objects, transport them to the target site, and eject cargos strapped onto magnetic nanorobot. DNA magnetic nanorobots based on a super-soft and super-elastic magnetic DNA hydrogel were able to exhibit shape-adaptive merits, and enhanced magnetically guided navigation velocity in limited and random spaces. Guided magnetic nanorobots are anticipated to provide elegant penetration in invasive medicine within the human body in the nearest future (as illustrated in **Figure 6**) [9].

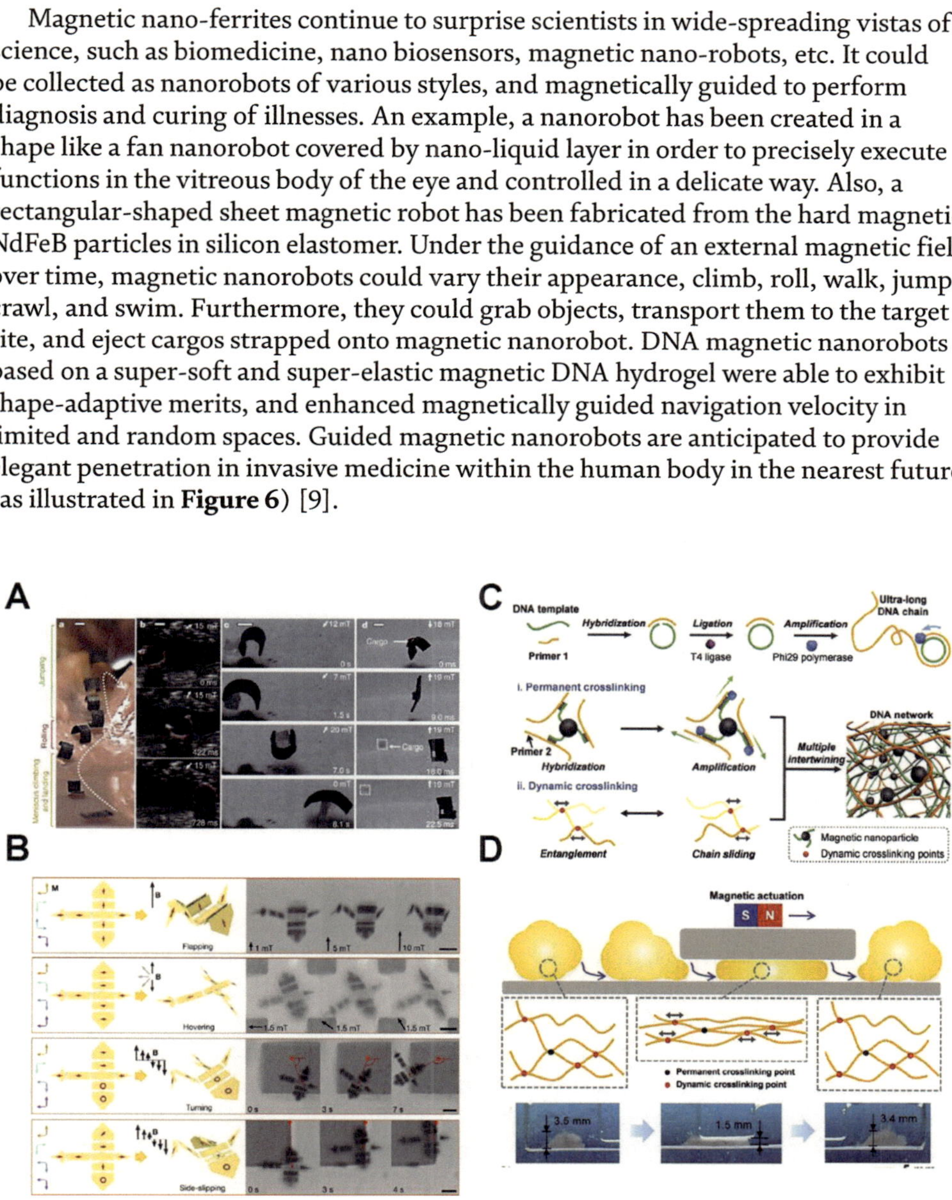

Figure 6.
(A) Small-scale soft-bodied robot in medical applications, (B) schematic and optical images of a microscale 'bird' mimicking four flying modes, (C) schematic of magnetic DNA hydrogel synthesis, and (D) shape adaptation characteristics of DNA robots [9].

2.6 Biosensors

Magnetic nanoparticles like nano-ferrite crystals continue to surprise scientists and the scientific community as they are used in enormous numbers of research, application, and biotechnological implementation vistas such as contrast agents in MRI, drug delivery, biosensors, hyperthermia, magnetic nanorobots, cancer cells killers, etc. Biosensors are responsive to biological signals emitted from certain cells or tissues which could be converted into electrical signals. Core–shell nanoparticles possess large surface areas and ultra-tiny particle sizes, which in turn, increases interaction areas and biosensor sensitivity. Magnetic nanoparticles should own some vital merits in order to be utilized as biosensors, such as extremely high chemical stability, high corrosion resistivity, simple and easy synthesis, high magnetic susceptibility, extremely high stability in physiological ambiance, and excellent dispersibility. Nano-ferrite core–shell nanoparticles are excellent candidates for biosensor implementations, imputing their unique structural, optical, magnetic, and piezoelectric merits. Nano-ferrites magnetic core is surrounded by inert shell to protect them from oxidation and make them inert while maintaining their magnetic demeanor. Gold shell covering Cobalt ferrite core is utilized as deoxyribonucleic acid (DNA) biosensor. Thiol-modified peptide nucleic acid oligomers attached to Au/Cobalt ferrite core-shell nanoparticles can interact with DNA target molecules. This modified sample is a good platform for the immobilization of biomolecules which can be used to detect point mutations or single nucleotide polymorphisms in DNA. Enzyme-free Glucose biosensors have been the main focus recently for the detection of diabetics imputing to their high thermal stability, extreme chemical stability, and reproducibility. Enzyme-free Glucose biosensors with Zinc ferrite/polypyrrole core–shell nanoparticles nanocomposites interestingly possess promoted electrochemical effectiveness toward Glucose oxidation with excellent sensibility. Copper ferrite/Polypyrrole core–shell nanoparticles are advanced nanomaterials utilized as Glucose sensors. These biosensors showed extremely surging electrochemical activity toward Glucose electro-oxidation. Polyvinyl pyrrolidone (PVP) capped Cobalt ferrite/Cadmium Selenide core–shell nano-composites were affirmed as excellent electrochemical biosensors for the detection of Rifampicin antibiotic. These new excellent biosensors possess long time span stability, extremely high sensibility, and reproducibility, which makes them excellent future candidates as a novel platform for the fabrication of electrochemical biosensors and further sensor implementations [23].

2.7 Brain stimulation

Magnetic ferrite nanoparticles continue to be very exciting nanocrystals imputing their unique structural, spectral, optical, chemical, physical, thermal, and especially electrochemical features. Then, they are excellent candidates for versatile implementations such as using them in brain stimulation. These nanocrystalline structures could be utilized in brain stimulation, which is established on the magneto-thermal effect under the influence of the alternating magnetic fields (AMF), whilst this strategy is known as the magneto-thermos-genetics strategy (magneto-genetics). A small comparison between the traditional procedures that are based on the utilization of chemicals and electrodes, and, the second-generation procedures that are based on the utilization of opto-genetics, and ultrasound, on the other hand, magneto-genetics has superiority over the first strategies. Magnetogenetics possess the superiority of cell type specificity as well as their superior temporal and spatial resolution. All and above Magneto-genetics methodology does not require tethering animals to an energy source [9]. Thence, scientists have an enormous interest in utilizing such nanostructured Magneto-genetics to activate the heat-sensitive transient receptor potential vanilloid 1 (TRPV1, ˃ 316 K) to attain further brain stimulation. Superparamagnetic nanocrystals including Manganese nano-ferrites, Magnetite (Fe3O4), Ferritin, Cobalt nano-ferrite core @ Manganese nano-ferrite core–shell nanostructure, have been utilized in the activation of thermos-sensitive TRPV1 channel. Manganese nano-ferrites showed that by heated up they could activate the TRPV1 channel. Then, by side some results happened, like, intracellular calcium concentration increased, triggered action potentials in primary hippocampal neurons, and elicited retracting motions in partially anesthetized *C. elegans* worms. Ferri-magnetic Magnetite (Fe_3O_4) nanoparticles coated with anti-His antibody could target the TRPV1 channel with an extracellular His × 6 epitope tag, which could remotely activate TRPV1 and stimulate the synthesis and release of proinsulin for regulation of insulin production in mice [9]. Green fluorescent protein (GFP) tagged ferritin nanoparticles have also been synthesized and intracellularly associated with a camelid anti-GFP-TRPV1, which could initiate calcium-dependent insulin transgene expression and modulate neuronal activities to induce feeding. A mutated chloride-permeable TRPV1 was further developed, and the neuronal inhibition was achieved via the same heat stimulation that responded to blood glucose levels and decreased feeding. Ferri-magnetic Magnetite (Fe_3O_4) could remotely activate TRPV1 and recall excitation in the targeted ventral tegmental area in vivo. Cobalt nano-ferrites @ Manganese nano-ferrites core–shell was early synthesized, however, with their superior magneto–thermal performances, they were utilized to stimulate neurocircuit and modulate behavior in awake mice, as illuminated in **Figure 7A**. The heat lost by superparamagnetic nano-ferrite particles could trigger the release of tiny molecules (agonist or inhibitor) from the thermally sensitive lipid vesicles and be applied to the thermally sensitive lipid vesicles and be applied to chemo-genetic activation of engineered receptors, as illuminated in **Figure 7B**. The chemo-genetic modulation of targeted neural circuits using superparamagnetic nano-ferrite particles provides a vital and novel strategy for brain stimulation so as to investigate brain functions and neurological diseases [9].

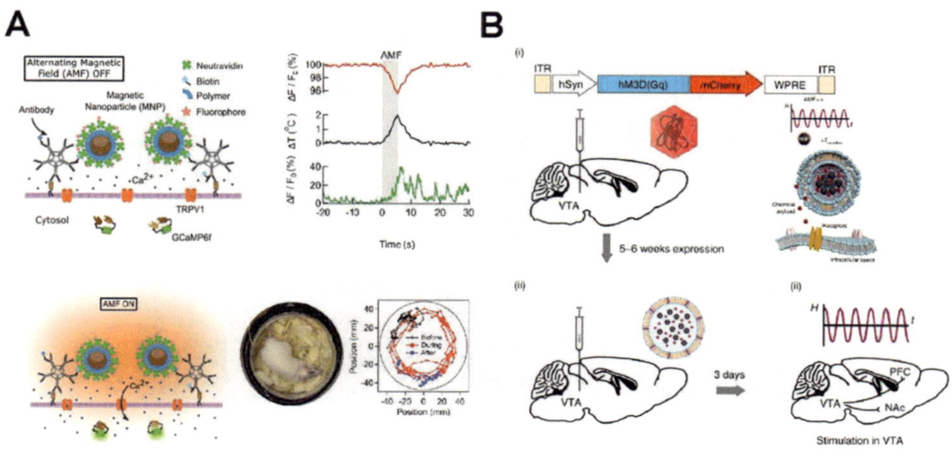

Figure 7.
(A) Magnetic nano-ferrite particles (MNFPs) under alternating magnetic field (AMF) activate TPRV1 channels, (B) chemo-magnetic stimulation in vivo [9].

2.8 Antibacterial applications

Nano-ferrite particles continue to be outstanding nanoparticles for versatile potential applications in numerous vistas including, biosensors, drug delivery, high-density information storage, magnetic refrigeration, etc. Among these, nano-spinels have been extensively investigated imputing their unique chemical and physical merits and their technological implementations in the biomedical vistas, antimicrobial, anti-fungi, photocatalysts, etc. Nanoscale substances like nano-ferrite possess almost similar sizes compared to biological molecules, so they can penetrate easily and interact with these biomolecules, and not possible for macromolecules. As these ultra-tiny nanoparticles were reduced to nano-scale extent they considerably changed their magnetic, optical, electrical, and chemical features, which also promoted these NPs to interact with biological systems in a unique strategy [26–28]. Aluminum Zinc nano-ferrite particles agglomeration influence has an important antibacterial activity, and fine dispersity of ultrafine nanoparticles, which is fundamental for efficient antibacterial activities. Antibacterial activity of Aluminum Zinc nano-ferrite particles obviously precisely occurs when prepared nanocrystals interact with cellular membrane resulting in production of Reactive Oxygen Species (ROS). The inhibition zone of *E. coli* and *Pseudomonas aeruginosa* against at Aluminum Zinc nano-ferrite particles and Nickel Zinc nano-ferrite particles is pictured and illuminated in **Figure 8**. Zinc nano-ferrite particles provided an antibacterial activity against Gram-negative with the maximum inhibition zone of 10 mm. For Aluminum Zinc nano-ferrite particles, they provided antibacterial activity against *Pseudomonas aeruginosa* with a maximum inhibition zone of 14 mm. Whilst, on the other hand, Nickel Zinc nano-ferrite particles provided antibacterial activity against E. coli with a supreme inhibition zone

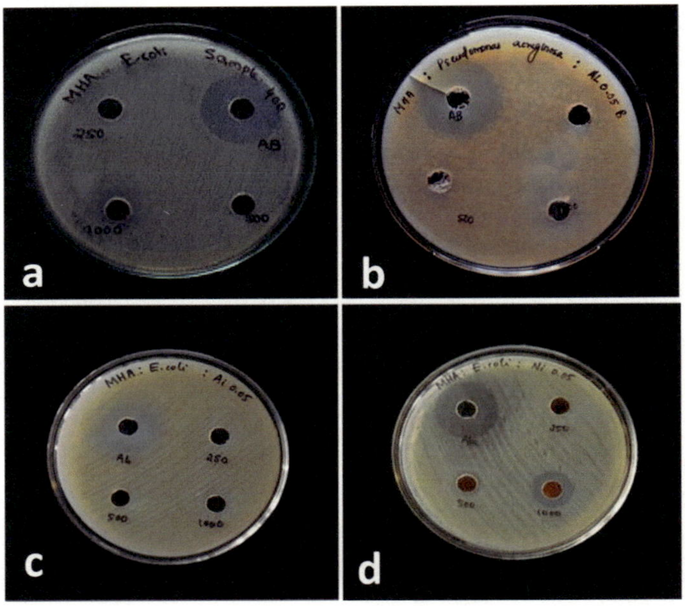

Figure 8.
Antibacterial activity by agar well-cut diffusion method against E. coli of (a) zinc nano-ferrite at 400°C, (b) antibacterial effect against Pseudomonas aeruginosa of aluminum zinc nano-ferrite (0.05), (c) antibacterial effect against E. coli of aluminum zinc nano-ferrite (0.05), (d) antibacterial effect against E. coli of nickel zinc nano-ferrite (0.05) [26].

of 17 mm. Antibacterial activity of Aluminum Zinc nano-ferrite particles essentially occurs in an accurate mechanism when ultrafine nanocrystals interact with cell membranes resulting in the increased production of ROS. The affirming evidence proposes that when the biological cells are exposed to the ultra-tiny nano-ferrite particles, this can produce an imbalance in the cellular reduction–oxidation process of bacteria by changing reactive oxygen species that may interrupt the bacterial-antioxidant defense responses. The increased ROS level generates oxidative stress, which is believed to be the contributing factor in the antimicrobial activity of NPs [26–28].

2.9 Anti-fungi applications

Superparamagnetic nano-ferrite particles have provided superior antibacterial features against various bacteria and fungi. Cobalt Ferrite nanoparticles are obviously utilized for biomedical implementations imputing their ability to combat various human infections. The ultra-fine size nature of these nano-ferrites as well as their high specific surface area makes them effective against harmful microorganisms [29]. The antifungal influence of Barium nano-ferrite particles has been precisely investigated as illuminated in **Figure 9**. The broth media of each condition was seeded in independent agar plates and the colony forming units (UFC) were assessed and referred over control (**Figure 9**).

2.10 Nano-ferrites for wound healing

Skin is the Extremely large body organ that spreads over the surface of the whole body. Skin including hair, nails, oil, and sweat glands form the human integumentary system. Skin main function is to defend the whole human body from external dangers such as; insects, parasites, bacteria, viruses, infections, wounds, sores, temperature, heat, humidity, chemicals, *etc*. Skin, the fundamental organ for humans, possesses esthetic nature as well as playing an active role in the immune system. Skin wounds essentially originate from accidents by sharp articles, metal fragments, sharp glass blades, or chronic ulcerations, thus it needs rapid therapeutic intervention or fast treatment and cure. Natural skin regeneration procedures may face several hurdles like bacterial infection, hence, hastening to heal is vital in skin regeneration healing.

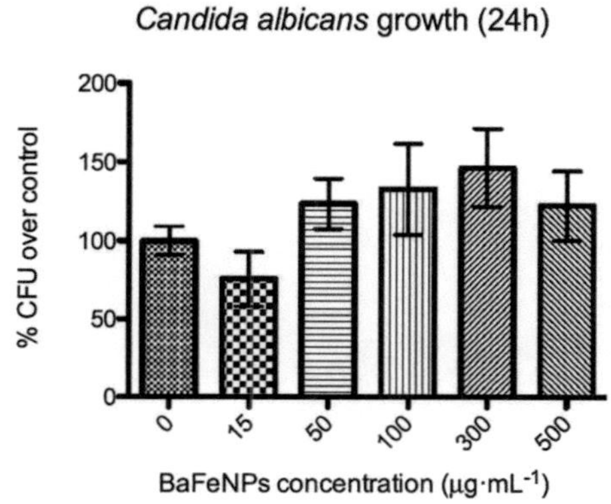

Figure 9.
Growth of Candida albicans treated with different concentrations of barium ferrite nanoparticles over control [29].

New synthesized nanostructured biomaterials must have several merits such as biocompatibility and biodegradability to avoid inflammation. Furthermore, they must have a special surface that assists cell adherence, splitting, and regeneration. Moreover, there should be a suitable porosity ratio that allows construction of blood vessels. Otherwise, diverse nanostructured biomaterials, such as bio-polymeric matrices (such as e-poly-caprolactone (PCL)/ampicillin fibrous nano-composition), have been proposed for enhancement of tissue integration [30–35].

Graphene oxide (GO) is a honeycomb crystalline structure formed of a single sheet or layer made up of a 2D array of Carbon atoms. GO is a new elegant and smart substance as it exhibits superior mechanical attributes and smart biocompatibility merits. It is obvious that GO possesses brilliant biological activity, and above this, the functional groups on GO surface, such as OH groups enhance hydrophilicity. There are several routes of synthesizing new intelligent nanocomposite structures in order to promote new smart structural, chemical, physical, thermal, and biological features. On the other hand, rare earth nanostructured crystals possess biological activity such as enhancing the therapeutic development in the new race of advancement of biomedical nanomaterials engineering. Thence, uniting nanostructured materials such as nano-ferrites, GO, rare earth nanostructures, etc., upon their unique functional demeanors could provide a delicate strategy in order to secure viable biomaterials in skin wound healing. Praseodymium oxide (Pr_6O_{11}), hematite (Fe_2O_3), graphene oxide (GO), and Polycaprolactone (PCL) based polymeric film nanocomposites (NCs)

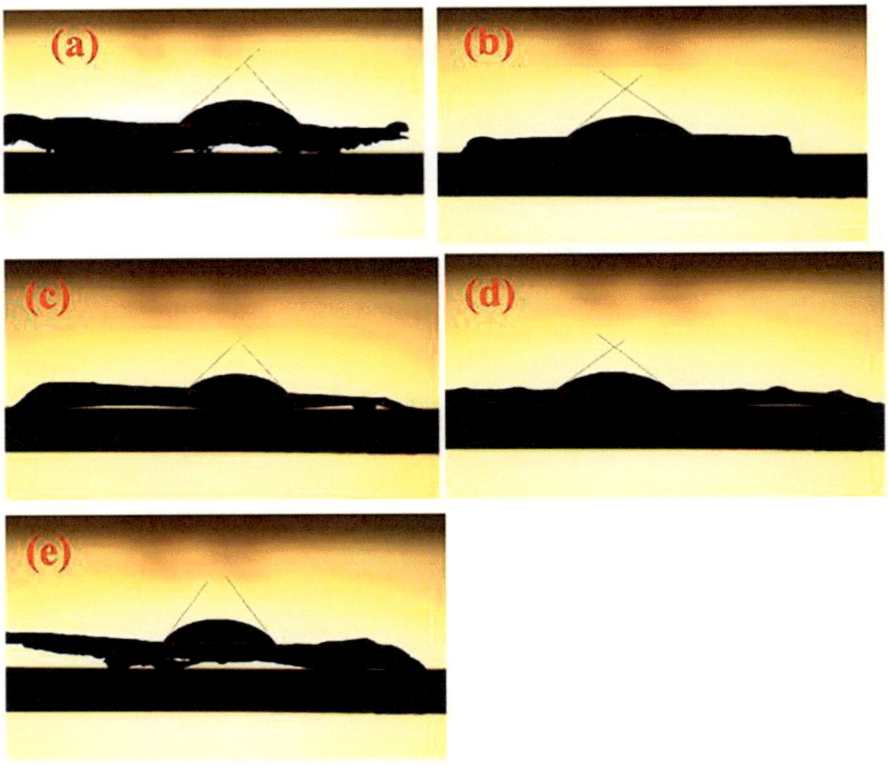

Figure 10.
Contact angle of e-poly-caprolactone (PCL) based fibrous nano-composite; where (a) PCl, (b) Pr_6O_{11}@PCl, (c) Fe_2O_3@PCl, (d) Pr_6O_{11}/Fe_2O_3@PCl, and (e) Pr_6O_{11}/Fe_2O_3/GO@PCl [31].

have been combined and synthesized to produce various polymeric nanocomposites for investigation of their biological activity and enhancement of skin wound healing. Pr_6O_{11}/Fe_2O_3/GO@PCL film NCs presented a promotion in skin wound healing at which it possesses a high capacity for healing, reaching around 80%, as it assists the division of normal cells promoting wound healing. Thence, these smart new nanocomposites could consolidate in production of new applicable marvelous nanostructures in wound healing applications [30–35].

The contact/wettability angle is a pointer to bio-applicability owing to the fact that the lesser the contact angle the higher the biological potential. The studied compositions' contact angle values according to **Figure 10** are ≈ 43, 38, 46, 38, and 55° for PCL, Pr_6O_{11}@PCL, Fe_2O_3@PCL, Pr_6O_{11}/Fe_2O_3@PCL, and Pr_6O_{11}/Fe_2O_3/GO@PCL, respectively. The displayed data support the Pr_6O_{11} effect on lessening contact angle, thus affecting nano-ferrites biological potential positively [30–35].

3. Conclusion

Nano-ferrites are ultra-tiny crystalline structures and possess versatile expanding applications ranging from permanent magnets to internet, computer, and mobile technologies. Above this, they possess very important biomedical implementations including, drug delivery, anti-cancer function, anti-fungi, anti-bacteria, cell signaling, hyperthermia, brain stimulation, magnetic resonance imaging (MRI), nanorobots, biosensors, wound healing, etc. Nano-ferrites showed great success in anticancer function towards various types of cancer cells, such as breast cancer as well as drug delivery and drug release. Their success in magnetic hyperthermia extends to cancer treatment, where nano-ferrites have been used under the influence of external magnetic fields to eliminate cancer cells and tumors. Nano-ferrites are used as magnetic resonance imaging (MRI) contrast agents. Their obvious utilization in the magnetic nanorobots and biosensors industries adds a new success to their usage. Also, scientists have utilized Manganese nano-ferrites to activate the heat-sensitive transient receptor potential vanilloid 1 (TRPV1, ˃ 316 K) to attain further brain stimulation, so as to investigate the brain functions and neurological diseases. Moreover, nano-ferrites have excellent antibacterial and antifungal activity. Pr_6O_{11}/Fe_2O_3/GO@PCL film nanocomposites have been used in skin wound healing treatment.

Author details

Amina Ibrahim Ghoneim
Faculty of Science, Physics Department, Tanta University, Tanta, Egypt

*Address all correspondence to: mona_ghoneim@yahoo.com;
amina.ali@science.tanta.edu.eg

IntechOpen

© 2023 The Author(s). Licensee IntechOpen. This chapter is distributed under the terms of the Creative Commons Attribution License (http://creativecommons.org/licenses/by/3.0), which permits unrestricted use, distribution, and reproduction in any medium, provided the original work is properly cited.

References

[1] Verma S, Chawla A, Pushkarna I, Singh A, Godara SK. Understanding the phase evolution with temperature in pure ($BaFe_{12}O_{19}$) and zinc-zirconium co-doped barium hexaferrite ($BaZnZrFe_{10}O_{19}$) samples using Pawley and Rietveld analysis. Materials Today Communications. 2021;**27**:102291. DOI: 10.1016/j.mtcomm.2021.102291

[2] Jiang X, Jia H, Wu C, Yu Z, Luo H. Cation distribution and magnetic characteristics of textured $BaFe_{12-x}Sc_xO_{19}$ hexaferrites: Experimental and theoretical evaluations. Journal of Alloys and Compounds. 2020;**835**:155202. DOI: 10.1016/j.jallcom.2020.155202

[3] Amer M, Meaz T, Attalah S, Ghoneim A. Structural and magnetic studies of Ti^{4+} substituted M-type $BaFe_{12}O_{19}$ hexa-nanoferrites. Materials Science in Semiconductor Processing. 2015;**40**:374-382. DOI: 10.1016/j.mssp.2015.07.007

[4] Vinnik D, Zhivulin V, Starikov A, Gudkova S, Trofimov E. Influence of titanium substitution on structure, magnetic and electric properties of barium hexaferrites $BaFe_{12-x}Ti_xO_{19}$. Journal of Magnetism and Magnetic Materials. 2020;**498**:166117. DOI: 10.1016/j.jmmm.2019.166117

[5] Goldman A. Modern Ferrite Technology. New York: Marcel Dekker Inc.; 1993

[6] Godara S, Kaur V, Chuchra K, Narang S, Singh G. Impact of Zn^{2+}-Zr^{4+} substitution on M-type barium strontium Hexaferrite's structural, surface morphology, dielectric and magnetic properties. Results in Physics. 2021;**22**:103892. DOI: 10.1016/j.rinp.2021.103892

[7] Thang P, Tiep N, Ho T, Co N, Hong N. Electronic structure and multiferroic properties of (Y, Mn)-doped barium hexaferrite compounds. Journal of Alloys and Compounds. 2021;**867**:158794. DOI: 10.1016/j.jallcom.2021.158794

[8] Kumar S, Guha S, Supriya S, Pradhan L, Kar M. Correlation between crystal structure parameters with magnetic and dielectric parameters of Cu-doped barium Hexaferrite. Journal of Magnetism and Magnetic Materials. 2020;**499**:166213. DOI: 10.1016/j.jmmm.2019.166213

[9] Wang Y, Miao Y, Li G, Su M, Chen X, Zhang H, et al. Engineering ferrite nanoparticles with enhanced magnetic response for advanced biomedical applications. Materials Today Advances. 2020;**8**:100119. DOI: 10.1016/j.mtadv.2020.100119

[10] Nagy Ľ, Zeleňáková A, Hrubovčák P, Barutiak M, Lisnichuk M, Bednarčík J, et al. The role of pH on the preparation of citric acid coated cobalt ferrite nanoparticles for biomedical applications. Journal of Alloys and Compounds. 2023;**960**:170833. DOI: 10.1016/j.jallcom.2023.170833

[11] Kefeni KK, Msagati TAM, Nkambule TTI, Mamba BB. Spinel ferrite nanoparticles and nanocomposites for biomedical applications and their toxicity. Materials Science & Engineering C. 2020;**107**:110314. DOI: 10.1016/j.msec.2019.110314

[12] Tangra AK, Singh S, Singh G. Investigation of the magnetic behavior of the ferrites of alkali and alkaline earth metals for biomedical application. Materials Today: Proceedings. 2021;**36**:621-625. DOI: 10.1016/j.matpr.2020.04.032

[13] HuiZhang JW, Zeng Y, Wang G, Han S, Yang Z, Li B, et al. Leucine-coated cobalt ferrite nanoparticles: Synthesis, characterization and potential biomedical applications for drug delivery. Physics Letters A. 2020;**384**:126600. DOI: 10.1016/j.physleta.2020.126600

[14] Shamima Nasrin F-U-Z, Chowdhury SMH. Study of hyperthermia temperature of manganese-substituted cobalt nano ferrites prepared by chemical co-precipitation method for biomedical application. Journal of Magnetism and Magnetic Materials. 2019;**479**:126-134. DOI: 10.1016/j.jmmm.2019.02.010

[15] Somvanshi SB, Kharat PB, Khedkar MV, Jadhav KM. Hydrophobic to hydrophilic surface transformation of nano-scale zinc ferrite via oleic acid coating: Magnetic hyperthermia study towards biomedical applications. Ceramics International. 2020;**46**:7642-7653. DOI: 10.1016/j.ceramint.2019.11.265

[16] Sanpo N, Berndt CC, Wen C, Wang J. Transition metal-substituted cobalt ferrite nanoparticles for biomedical applications. Acta Biomaterialia. 2013;**9**:5830. DOI: 10.1016/j.actbio.2012.10.037

[17] Kalaiselvan CR, Laha SS, Somvanshi SB, Tabish TA, Thorat ND, Sahu NK. Manganese ferrite (MnFe2O4) nanostructures for cancer theranostics. Coordination Chemistry Reviews. 2022;**473**:214809. DOI: 10.1016/j.ccr.2022.214809

[18] Bentarhlia N, Elansary M, Belaiche M, Mouhib Y, Lemine OM, Zaher H, et al. Evaluating of novel Mn–Mg–Co ferrite nanoparticles for biomedical applications: From synthesis to biological activities. Ceramics International. 2023. DOI: 10.1016/j.ceramint.2023.10.017 (In Press)

[19] Nasrin S, Hasan M, Khurshida Sharmin MA, Islam AN, Hottori H, Tanak AKM, et al. Nanocrystalline Mn-doped Ni–Cu ferrites with a high cut-off frequency and initial permeability: Suitable for advanced electronic devices and biomedical applications. Materials Chemistry and Physics. 2023;**297**:127322. DOI: 10.1016/j.matchemphys.2023.127322

[20] Ahmad N, Alomar SY, Albalawi F, Khan MR, Farshori NN, Wahab R, et al. Anticancer potentiality of green synthesized Mg-Co ferrites nanoparticles against human breast cancer MCF-7 cells. Journal of King Saud University – Science. 2023;**35**:102708. DOI: 10.1016/j.jksus.2023.102708

[21] Nigam A, Pawar SJ. Structural, magnetic, and antimicrobial properties of zinc doped magnesium ferrite for drug delivery applications. Ceramics International. 2020;**46**:4058-4064. DOI: 10.1016/j.ceramint.2019.10.243

[22] De D, Upadhyay P, Das A, Ghosh A, Adhikary A, Goswami MM. Studies on cancer cell death through delivery of dopamine as anti-cancer drug by a newly functionalized cobalt ferrite nano-carrier. Colloids and Surfaces A: Physicochemical and Engineering Aspects. 2021;**627**:127202. DOI: 10.1016/j.colsurfa.2021.127202

[23] Kurian M, Thankachan S. Structural diversity and applications of spinel ferrite core—Shell nanostructures—A review. Open Ceramics. 2021;**8**:100179. DOI: 10.1016/j.oceram.2021.100179

[24] Hatamie S, Balasi ZM, Ahadian MM, Mortezazadeh T, Shams F, Hosseinzadeh S. Hyperthermia of breast cancer tumor using graphene oxide-cobalt ferrite magnetic nanoparticles in mice. Journal of Drug Delivery Science and Technology. 2021;**65**:102680. DOI: 10.1016/j.jddst.2021.102680

[25] Arshad JM, Raza W, Amin N, Khalid Nadeem M, Imran Arshad M, Khan A. Synthesis and characterization of cobalt ferrites as MRI contrast agent. Materials Today: Proceedings. 2021;**47**:S50-S54. DOI: 10.1016/j.matpr.2020.04.746

[26] Jyothish B. John Jacob, synthesis and characterization of Ni^{2+} and Al^{3+} doped zinc ferrite nanoparticles for antibacterial, antioxidant, and anticancer (MCF-7) analysis. Chemical Physics Impact. 2023;**6**:100209. DOI: 10.1016/j.chphi.2023.100209

[27] ElHajjar I, Al Bitar M, Zahr S, Khalil M, Awad R. Investigation of the physical properties and antibacterial activity of various ferrite, chromite, and aluminate nanocomposites. Journal of Alloys and Compounds. 2023;**968**:171953. DOI: 10.1016/j.jallcom.2023.171953

[28] Kumar R, Jasrotia R, et al. A new hybrid non-aqueous approach for the development of Co doped Ni-Zn ferrite nanoparticles for practical applications: Cation distribution, magnetic and antibacterial studies. Inorganic Chemistry Communications. 2023;**157**:111355. DOI: 10.1016/j.inoche.2023.111355

[29] Alvino L, Pacheco-Herrero M, López-Lorente ÁI, Quiñones Z, Cárdenas S, González-Sánchez ZI. Toxicity evaluation of barium ferrite nanoparticles in bacteria, yeast and nematode. Chemosphere. 2020;**254**:126786. DOI: 10.1016/j.chemosphere.2020.126786

[30] Qi W, Jin L, Cuixi W, Liao H, Zhang M, Zhu Z, et al. Treatment with FAP-targeted zinc ferrite nanoparticles for rheumatoid arthritis by inducing endoplasmic reticulum stress and mitochondrial damage. Materials Today Bio. 2023;**21**:100702. DOI: 10.1016/j.mtbio.2023.100702

[31] Zrieq R, Alzain MA, Haouas N, Ali RM, Elabbasy MT, El-Morsy MA, et al. Surface modification of praseodymium oxide/hematite doped into Polycaprolactone for enhanced wound management demands. Journal of Saudi Chemical Society. 2023;**27**:101708. DOI: 10.1016/j.jscs.2023.101708

[32] Zhang J, Gao X, Ma D, He S, Bingwen D, Yang W, et al. Copper ferrite heterojunction coatings empower polyetheretherketone implant with multi-modal bactericidal functions and boosted osteogenicity through synergistic photo/Fenton-therapy. Chemical Engineering Journal. 2021;**422**:130094. DOI: 10.1016/j.cej.2021.130094

[33] Farzaneh S, Hosseinzadeh S, Samanipour R, Hatamie S, Ranjbari J, Khojasteh A. Fabrication and characterization of cobalt ferrite magnetic hydrogel combined with static magnetic field as a potential bio-composite for bone tissue engineering. Journal of Drug Delivery Science and Technology. 2021;**64**:102525. DOI: 10.1016/j.jddst.2021.102525

[34] Naresh U, Jeevan Kumar R, Naidu KCB. Hydrothermal synthesis of barium copper ferrite nanoparticles: Nanofiber formation, optical, and magnetic properties. Materials Chemistry and Physics. 2019;**236**:121807. DOI: 10.1016/j.matchemphys.2019.121807

[35] Ameen F, Majrashi N. Recent trends in the use of cobalt ferrite nanoparticles as an antimicrobial agent for disability infections: A review. Inorganic Chemistry Communications. 2023;**156**:111187. DOI: 10.1016/j.inoche.2023.111187

Chapter 3

Functionalized Ferrites for Therapeutics and Environmental Pollution Management

Tonmoye Sarkar Shathi and Abdur Rahman

Abstract

Surface-functionalized ferrite materials are the ultimate products obtained from micro/nanofabrication of one or more metal containing magnetic iron-based oxides and their surface fine-tuning with suitable molecules for desired applications. Appropriate functionalization of ferrite surface often implants a wide range of application-specific physicochemical characteristics. Herein, we have discussed surface functionalization of ferrites with different organic molecules, inorganic oxides, metals, and polymeric materials. Impacts of surface functionalization on the dispersibility, biocompatibility, conductivity, photocatalytic activity, and pH responsiveness of ferrite particles and their therapeutic and environmental potentials are also highlighted. Then, some widely used and important functionalization routes like coupling, ligand exchange, polymer encapsulation, and sol-gel techniques are illustrated. Finally, a brief overview of biomedical and environmental pollutant mitigation efficacies of the functionalized ferrite compounds is emphasized.

Keywords: surface functionalization, colloidal stability, ferrites, biomedicine, pollutant mitigation

1. Introduction

Ferrites are ferrimagnetic iron oxide-based materials consisting of a small portion of one or more other metallic elements such as Ba, Sr., Mn, Bi, Co, Ni, and Zn. Low-cost, easy formulation, and magnetism of ferrites make them a potential candidate for technological and biomedical applications. However, poor colloidal stability, high band gap energy, and absence of suitable surface functionality highly limit their practical applications. To address these issues, various strategies have been applied. Among them, surface functionalization of ferrites is one of the important methods, which offers better dispersion stability, magnetism, high processability, desired functionality for anchoring on demand molecules, and relatively lower band gap energy to ease photocatalytic reactions. Functionalized ferrites are widely studied materials in the fields of energy, electronics, magnetics, catalysis, and biomedical technology for the last several decades. Here, the functionalization strategies of ferrites and application potentials of functionalized ferrites in therapeutic delivery and catalytic degradation of different environmental pollutants will be illustrated based on the recent literature.

Recently, many technological advancements have been made by using ferrite materials in the fields of biomedical sciences and environmental pollutant remediation engineering. Ferrites are a distinct class of magnetic nanoparticles (MNPs) having the general formula of MFe_2O_4, where the tetrahedral cationic site is occupied by a divalent cation, M^{2+}, and the trivalent Fe^{3+} ions occupy the octahedral cationic sites of the crystal lattice [1]. Ferrite NPs possess high surface energy-to-volume ratio and excellent superparamagnetic characteristics. Ferrites materials can be synthesized using various methods namely coprecipitation [2], thermal decomposition [3], microemulsion [4], hydrothermal [5], and sol-gel process [6]. Ferrites are often used as strong magnetic adsorbents, sensors, payload carriers, and imaging contrast agents. However, these practical applications inevitably hindered due to the colloidal instability of ferrites in the dispersing medium [7]. In such cases, appropriate surface modification of ferrite materials can prevent excessive aggregation and immature leaching from the reacting environment while enhancing or retaining their core magnetic responsiveness [1]. In this chapter, our focus is to summarize some ferrite-functionalizing materials and their synthesis routes keeping mainly their therapeutic and environmental pollution management applications in mind.

Some of the ferrite particles are inevitable in the field of therapeutic applications like nanocarrier for payloads, biosensor for disease diagnosis, hyperthermia [8], and environmental hazard materials management *via* separation by adsorption and photodegradation of pollutants [9]. Unique saturation magnetization, permeability, and structural anisotropic properties of MNPs make them quite desirable materials to the researchers for their widespread usages in biomedical and technological fields [10]. However, these properties are quite insufficient for effective usage, especially in the biomedical and environmental areas. Basically, fine-tuning the surface properties of the ferrite nanoparticles namely colloidal stability, aqueous dispersibility, biocompatibility and nontoxicity are the key properties for their reactivity towards the surrounding environments. These properties are also crucial for determining their application potentials in different fields [11]. Functionalization is a vital technique for tailoring the surface properties of nanoparticles with different organic and inorganic nanoscale materials through the covalent and noncovalent bonds [12]. The aim of functionalization is to enhance the physicochemical and biological characteristics of the resulting ferrites.

2. Widely used materials for surface functionalization

Surface functionalization of the magnetic ferrite nanoparticles serves the advantages of site specificity, enhanced colloidal stability, and biocompatibility [8]. A wide range of applications of ferrites has been achieved *via* the controlled tuning of their surface properties. Nevertheless, the aftermath of the modification process is severely controlled by the physical, chemical, and electrical properties of the incoming ligands, dopants, and coating materials. Hence, it is important to have an insight into some of the readily used functionalizing materials and their influences on the properties of ferrite particles. **Figure 1** illustrates the classification of different functionalizing materials for the surface modification of ferrite NPs.

2.1 Molecular organic compounds

There are many small organic molecules that are commonly used as stabilizing and functionalizing agents like amines and thiols for modification of the surface

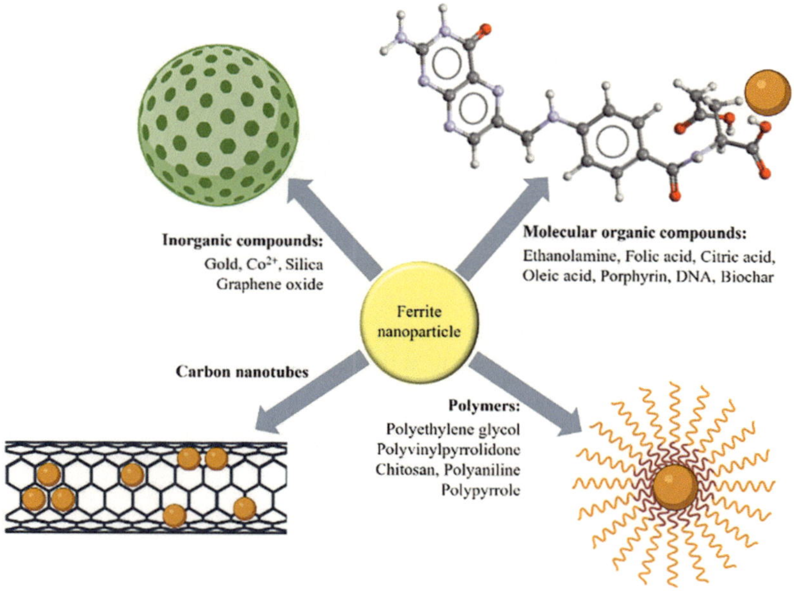

Figure 1.
Different materials used for ferrite functionalization.

of ferrites. One such example is ethanolamine, which can modify the ferrite NP surfaces in a simple one-step process [13, 14]. The surface amine groups lower the hydrodynamic diameter (HDD) of the functionalized $CoFe_2O_4$ MNPs and facilitate their usage in biomedical applications [13]. The hardly basic nature of -NH_2 groups of $MgFe_2O_4$-NH_2 adsorbent helps in the selective chemisorption of different heavy metal ions (Pb^{2+}, Cu^{2+}, and Zn^{2+}) *via* a complexation reaction [14]. For thiol functionalization, a mixture of tetraethyl orthosilicate (TEOS) and (3-mercaptopropyl)-trimethoxysilan is much desired by many researchers [15, 16]. It includes the silica coating of the ferrite core to which the -SH group is attached. Like amine functionalization, the heavy metal adsorption process of thiol-functionalized ferrites is influenced by the Lewis acid-base interactions [15].

Folic acid (FA) is a poorly water-soluble biologically important molecule. Its overexpression in the vast majority of cancer cells gives the FA-modified nanoparticles a better chance for internalization into the body and enhances the efficacy of targeted delivery of drug molecules. FA functionalization facilitates the pH-responsive release of the cationic anticancer drug doxorubicin (DOX) from cobalt ferrite magnetic nanoparticles (CFMNPs). At basic pH, FA is converted into folate ions and immobilizes a higher amount of DOX molecules. The acidic environment of cancer cells stimulates the breaking of this electrostatic interaction and enables the burst release of the DOX molecules [8]. FA is widely used for rendering cancer cell targetability to different anticancer drug nanocarriers [8, 17, 18]. Citric acid has been readily used in the synthesis of biocompatible magnetic fluids (BMFs) for enhanced colloidal stability of the nanoparticles [19, 20]. Citrate-coated cobalt (cit-CF) and nickel ferrite (cit-NF) nanoparticles exhibit a dose-dependent radiation cytotoxicity against MCF-7 cancerous cells and are considered a suitable radiosensitizer for cancer treatment [21]. Oleic acid (OA) is a type of fatty acid that is used as a surfactant and forms a dense protective shell layer around the nanoparticle surface during its functionalization process. In addition to its ambiphilic nature, OA imposes a non-immunogenic and anti-inflammatory environment on the attached nanoparticle. Cell viability studies by Sandeep *et al.* showed that only a minimum dose rate (8 mg/ml) of the OA-coated zinc ferrite (OA-ZNF) is required for effective hyperthermia therapy [22]. For effective biological applications, Nam et al. transferred the OA/oleylamine (OLA)-stabilized $CoFe_2O_4$ particles from hydrophobic to hydrophilic phase through the poly(maleic anhydride-alt-1-octadecene) (PMAO) encapsulation [23]. Porphyrin, an organic heterocyclic compound, is a promising photoredox catalyst and photosensitizer. Functionalization by porphyrins aids the photocatalytic activities of ferrite NPs due to the ease of electron transfer through their highly π-conjugated systems [24]. Parnian *et al.* demonstrated that a porphyrin derivative, meso-tetraphenylporphine-4,4′,4″,4‴-tetracarboxylic acid (TCPP), can significantly enhance the photocatalytic activity of polythiophene-coated $ZnFe_2O_4$ (TCPP/$ZnFe_2O_4$@PTh) nanocomposite [25].

Monodispersed $MnFe_2O_4$ NPs were functionalized with biotin and single-stranded DNA (ssDNA) for effective detection of protein or complementary ssDNA-patterned substrate. Here, the low magnetocrystalline anisotropy and a high moment of the NPs helped the bio-functionalization and *in situ* magnetic detection at room temperature in both *in vitro* and *in vivo* biological environments [26]. Biochar (pyrogenic black carbon) is recognized as a cost-effective adsorbent for environmental remediation due to its unique surface porosity and functionality. Excessive phosphate can seriously endanger aquatic bodies. Superparamagnetic magnesium ferrite ($MgFe_2O_4$)/biochar magnetic composites (MFB-MCs) remove phosphate from an aqueous solution through an inner-sphere complexation mechanism [27].

2.2 Widely used polymers

Polyglycerols (PG), also known as polyglycidols, are remarkably attractive candidates for surface functionalization due to their voluminous hydroxyl groups and flexible polyether backbone. Rimesh *et al.* used L-α-phosphatidylethanolamine (PE) block as a hyperbranched polyglycidol lipopolymer to provide water solubility and biocompatibility to hydrophobic oleylamine (OA)-stabilized manganese ferrite ($MnFe_2O_4$@OA) nanoparticles [17]. Debarati *et al.* functionalized dopamine-bound cobalt-ferrite (CF-DA) nanoparticles with polyethylene glycol (PEG). PEG functionalization prevented the early dissolution of DA in physiological conditions and facilitated its controlled release into the cancer cells [28]. Polyvinylpyrrolidone (PVP) is another well-known nanoparticle stabilizer and dispersant. Most of the bare ferrites lose their biological uses and superparamagnetic nature due to the excessive aggregation in the solution. PVP coating resists the formation of aggregated clusters of cobalt ferrites ($CoFe_2O_4$) making monodispersed and long-circulating MNPs. The as-prepared PVP-$CoFe_2O_4$ showcased lower cytotoxicity during the MTT assay [29]. Sahira *et al.* showed that PVP-induced biocompatibility increased the cellular uptake ability of $MnZnFe_2O_4$ nanoparticles [30]. Jaberolansar *et al.* proved that the non-magnetic PVP matrix effectively handles the heat generated from the $Co_{0.3}Zn_{0.7}Fe_2O_4$ ferrite powder and enhances its usability for hyperthermia application [31].

The admirable film-forming ability, high transmembrane permeability, mechanical strength, nontoxicity, and biocompatibility of chitosan (CH) make it a useful nanoparticle stabilizing and functionalizing agent to attain a wide variety of applications. CH-coated $CoFe_2O_4$ nanocomposite can successfully immobilize horseradish peroxidase (HRP) for the effective detection of hydrogen peroxide (H_2O_2). Comparative studies with several existing studies proved that CH/$CoFe_2O_4$

electrodes exhibit larger liner range (3×10^{-2} to 8 mM), shorter response time (4 s), and enhanced sensitivity (23 nA/mM) [32]. Chitosan-functionalized $CaFe_2O_4$ MNPs possess 88.2% immobilization efficiency for ampicillin, which is mainly attributed to the electrophilic interactions of the protonated amino groups of CH moiety with ampicillin [33]. Moreover, Datna *et al.* reported that CH-coated $CoFe_2O_4$ MNPs exhibit stronger antimicrobial activity for both the Gram-negative *P. aeruginosa* and Gram-positive *E. faecalis* and *E. coli* than the uncoated ones [34]. Conducting polymers; such as polyaniline (PANI) and polypyrrole (PPy) comprise a useful group of materials for ferrite functionalization due to their tunable electrical properties and high physical flexibility. Experiments showed that the diamagnetic PANI coating enhances the electromagnetic shielding ability of MnZn ferrite (MZF) and NiMnZn ferrite (NiMZF) [35]. Sadeghinia *et al.* showed that the PANI filaments in the PANI/perlite-barium ferrite nanoparticles (PANI/PBF-NPs) cause a decrease in their surface areas and an increase in the pore volume. Thus, PANI contributes to the improvement of the electrical charge storage capability of the composite [36]. The extended p-conjugation of PANI with single- and double-bond alteration is responsible for its semiconducting nature that can facilitate the electrochemical properties of different spinel ferrites [37, 38]. Yan *et al.* showed that due to the dielectric loss of PANI/PPY coating, the saturation magnetization of the PANI(PPY)-$BaFe_{12}O_{19}/Ni_{0.8}Zn_{0.2}Fe_2O_4$ ferrite decreases while increasing its electrical conductivity [39].

2.3 Metallic doping

Substituting paramagnetic Co^{2+} ions with diamagnetic Zn^{2+} in the $CoFe_2O_4$ lattice profoundly impacts the magnetic aspects of the $Co_{1-x}Zn_xFe_2O_4$ compound. The as-synthesized mixed ferrite exhibits increased saturation magnetization and reduced anisotropy constant due to the dilution of spin moments by the Zn^{2+} ions [40]. Zn doping essentially improves the photocatalytic efficiency of ferrite NPs by lowering the band gap and reducing the recombination of photogenerated electrons and holes [41, 42]. Gold nanoparticles (Au NPs) possess rich surface chemistry for facilitating the healthcare-related application sectors. Au coating of ferrite NPs hinders the oxidation of the magnetic core and acts as a convenient platform for further surface functionalization. Juan *et al.* fabricated glyco-ferrites for MRI contrast agent application using bimetallic superparamagnetic XFe_2O_4@Au (X = Fe, Mn, and Co) nanocrystals. During the synthesis process, the Au shell forms stable thiol-Au bonds with the neoglycoconjugates [43]. Au NPs render biocompatibility along with tunable plasmonic characteristics to the superparamagnetic $MgFe_2O_4$. Such optical properties allow the easy detection of tissue and blood in the near-infrared (NIR) region for potential hyperthermia and drug delivery application [44].

2.4 Inorganic compounds

Cadmium sulfide (CdS)-coated ferrite nanocomposites are known to exhibit good photocatalytic activity for wastewater treatment. CdS is an n-type semiconductor with a large direct band gap under visible light irradiation. Ferrite nanoparticles improve the catalytic activity, anti-photocorrosion, recovery, and reuse of CdS catalyst. Together, this semiconductor-based photocatalysis formulates a green technology for environmental protection by degrading several organic dyes (e.g., 4-chlorophenol (4-CP), methylene blue (MB), Rhodamine B (RhB), and methyl orange (MO)) [45, 46]. Functionalization by the inert silica layer can screen magnetic dipole interactions between magnetic nanoparticles and in that way facilitate their aqueous dispersibility and biocompatibility for biomedical and bioengineering applications. The presence of abundant silanol groups at the silica-coated ferrite NP surfaces enables the scope for easy processability for multifunctional nanocomposite synthesis [47]. Banalata *et al*. tailored the mesoporous silica-coated superparamagnetic manganese ferrite (MSN) with 3-aminopropyl triethoxysilane (APTES) for amine functionalization, which is schematically shown in **Figure 2**. Next, the H_2N-MSN particles were conjugated with FA for targeted delivery of the anticancer drug, doxorubicin (DOX). The residual amine groups on the FA-MSN were labeled with the fluorescent dye rhodamine-B-isothiocyanate (RITC) for their cell internalization and detection. Hence, this functionalized entity can serve for cancer diagnosis and treatment [47].

Ferrite-decorated graphene oxide (GO) nanocomposites are well dispersed in aqueous/physiological media and biocompatible in nature due to the high density of oxygen-containing groups, such as carboxylic, hydroxyl, and epoxide groups on the surfaces and edges of GO. The unique mechanical, electrical, and surface characteristics

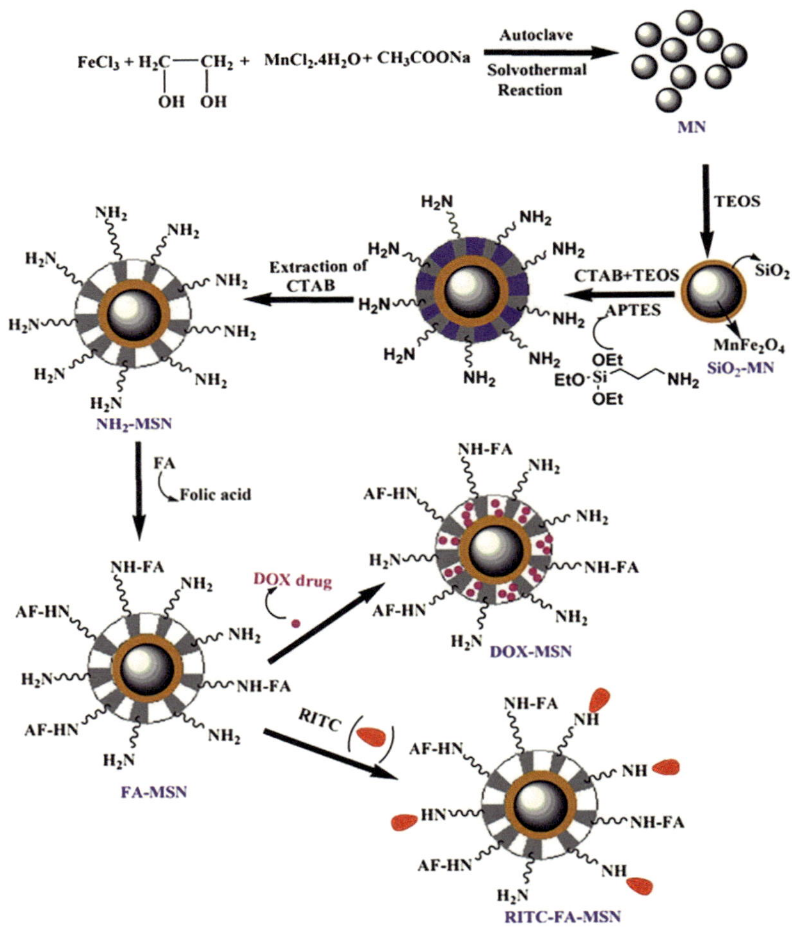

Figure 2.
Step-by-step synthesis of silica-coated manganese ferrite for DOX delivery (MSN). Copyright: Journal of colloid and Interface science [47].

of GO nanosheet promote many theranostic applications [48]. Yan *et al.* demonstrated that GO/$MnFe_2O_4$ nanohybrids showcase exceptionally high loading capacity for DOX mainly due to the strong π–π stacking and hydrophobic interaction between the hexagonal lattice of GO and the extended aromatic chain of DOX. Furthermore, the drug release mechanism is controlled by the pH and NIR irradiation [49].

2.5 Carbon nanotubes (CNTs)

Ferrite NPs decorated on CNTs possess excellent electrical, thermal, and mechanical properties. $CoFe_2O_4$ modified acid-functionalized multiwalled carbon nanotubes (MWCNT-COOH) are effective RhB adsorbent. Increasing MWCNT-COOH content from 29 to 75% increases the presence of active sites for electrostatic interactions and eventually increases the adsorption capacity from 5.165 to 42.68 mg g^{-1} [50]. On the other hand, Huixia *et al.* demonstrated the biomedical applications of MWCNT/$CoFe_2O_4$ nanocomposite. The strong supramolecular π–π stacking interaction between DOX and the side walls of CNTs results in its high loading capacity (about 75.2%). Moreover, a high T_2 relaxivity, low cytotoxicity, and pH-responsive drug release ability showcase the nanocomposite's potential as synergistic cancer diagnostic and chemotherapeutic agent [51]. Yan *et al.* coated the magnetoelectric multiwall carbon nanotubes (MWCNTs)/Fe–Ni alloy/NF particles with 3,4-ethylenedioxythiophene and pyrrole (PPy-PEDOT) copolymers in a complex core–shell structure. The conducting network of MWCNTs and PPy-PEDOT nanofiller improves the impedance matching and interfacial polarization, making a promising microwave absorber for electromagnetic pollution remediation [52].

3. Common functionalization strategies for the surface functionalization of ferrite materials

3.1 Coupling method

Nanoparticle functionalization *via* coupling reaction always occurs in the presence of a common intermediate to facilitate the energy transfer from one end of the reaction to another. Chaitali *et al.* cingulated FA on the CFMNP surfaces after the activation of FA using 1-Ethyl-3-(3-dimethylaminopropyl)-carbodiimide (EDC) and N-hydroxysuccinimide (NHS) in a dimethyl sulfoxide (DMSO) dispersion [8]. FA molecules were conjugated with different natural cytotoxic drug-encapsulated bovine serum albumin-calcium ferrite nanoparticles (BSA-CFNP) by the EDC coupling method [18].

3.2 Ligand exchange reaction

It is a post-synthesis surface modification method where the ligands of a preformed material are exchanged with another compound. Rimesh *et al.* utilized this process to exchange OA molecules from the MnFe2O4@OA nanoparticles with FA-modified PE hyperbranched polyglycidol (PE-HBPG-FA) [17]. Similarly, Seung *et al.* employed the ligand exchange method for nonpolar to polar (water) solvent phase transfer of manganese ferrites ($MnFe_2O_4$) dispersed in hexane into single-layered poly(ethylene glycol)-coated $MnFe_2O_4$. After that, the nano dispersion was converted into double-layered oleylphosphate(OA)-functionalized $MnFe_2O_4$ nanocrystals *via* ligand encapsulation [53].

3.3 Polymer encapsulation

Another useful post-synthesis functionalization process is polymer encapsulation. In general terms, encapsulation refers to the process in which a bioactive material is enclosed in an inert matrix mainly for stabilization purposes [54]. Common polymers like PVP, PEG, and CH coating on ferrite molecules are fabricated by simple mixing of the reagents followed by the application of ultrasonication or continuous stirring [28, 30, 34]. Others occur via *in situ* polymerization method initiated by heat, radiation or other materials that can generate the radical reactions. Ammonium persulfate (APS) initiated aniline polymerization reaction in the presence of powdered $CuFe_2O_4$ in order to yield highly crystalline PANI/CuFe2O4 nanocomposite [37]. Polymethacrylate (PMAA) functionalization of nanocrystalline nickel ferrites (NiFe2O4) was carried out by potassium persulfate ($K_2S_2O_8$)-initiated polymerization reaction in an acidic reaction medium (pH ~ 3) [55]. Magnetoelectric core-shell nanocarrier for chemotherapeutic drug methotrexate was fabricated by functionalizing $CoFe_2O_4$–$BaTiO_3$, $CoFe_2O_4$–Bi4Ti_3O_{12}, and Fe_3O_4–$BaTiO_3$ with PNIPAm [56].

3.4 Solution casting method

This process is used in thin-film formation mainly for the uniform distribution of the functionalizing materials in the polymer matrix. It is a convenient laboratory-friendly process in which polymer film adhesion with the reinforcing material (i.e., ferrite NPs in this case) is achieved by dispersing the polymer and NPs in a common solvent medium. Next, the evaporation of the solvent leaves out the desired polymer-functionalized nanocomposite [57]. The inclusion of multicomponent $CuFe_2O_4$/Cu_2O/CuO NPs in the polymethyl methacrylate (PMMA) matrix was carried out for enhanced antibacterial properties. The NPs-PMMA dispersion in acetone solution cast on a slide resulted in an antibiofilm nanocomposite [58]. Jay *et al.* synthesized CH-functionalized nanostructured $NiFe_2O_4$ (n-$NiFe_2O_4$-CH) thin film onto the indium tin oxide (ITO) glass substrate *via* this simple method [59].

3.5 Sol: Gel coating

Also known as inorganic sol-gel coating that occurs *via* successive hydrolysis and polycondensation of the precursor material (sol) into a three-dimensional continuous network (gel). This method is preferable due to better control over nanoparticle size and stability and coating homogeneity [60]. Ashkan *et al.* synthesized zinc silicate-coated superparamagnetic zinc ferrite composite using the sol-gel method. In this work, a simple tuning of pH value into the basic medium of the reacting solution paved the way for Zn^{2+} and Si^{4+} ion adsorption onto the cetyltrimethylammonium bromide (CTAB)-modified surfaces of $ZnFe_2O_4$ nanoparticles. After that, the intermediate CTAB layer was eliminated with the help of ethanol solution yielding zinc silicate shell over the $ZnFe_2O_4$ core. **Figure 3** shows the rough surface texture and microspherical size (100–300 nm) of the synthesized zinc silicate-$ZnFe_2O_4$ composite [61].

The sol-gel technique was utilized to yield a uniform coating of nanocrystalline TiO_2 shell around the $CoFe_2O_4$ MNPs core. Wuyou *et al.* induced the heterogeneous nucleation of TiO_2 *via* a slow and gradual supersaturation of $Ti(OC_3H_7)_4$ ethanol in 1:10 water–ethanol suspension of $CoFe_2O_4$ [62].

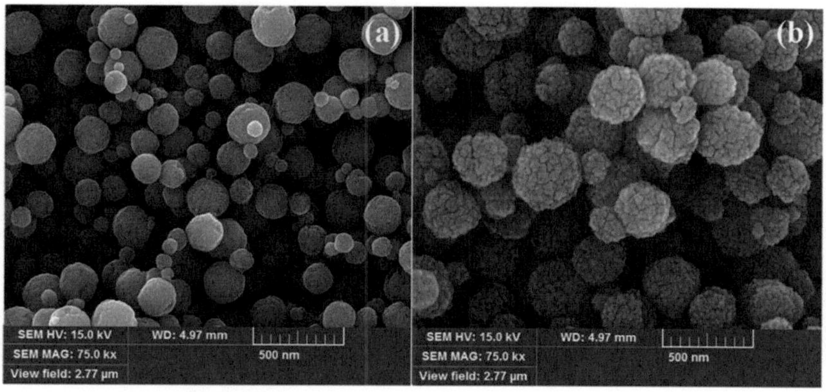

Figure 3.
The FESEM micrographs of $ZnFe_2O_4$ (a) and zinc silicate-$ZnFe_2O_4$ (b). Copyright: Ceramics international [61].

3.6 Stöber method

The Stőber method is a facile process for uniform silica coating on a nanoparticle surface through a sol-gel strategy. In simple words, ammonia-mediated hydrolysis of tetraethoxysilane (TEOS) yields silanol monomers at the nanoparticle surface. After that, the neighboring silanol monomers are condensed into a siloxane network cluster [63]. Likewise, Kooti *et al.* utilized this method to coat $CoFe_2O_4$ with SiO_2. The inert silica layer was later used for anchoring a molybdenum Schiff base onto the complex surface in order to enhance the catalytic activity for the oxidation of alkene. **Figure 4** shows the step-by-step reactions that occurred during the functionalization process [64].

4. Therapeutic applications of functionalized ferrites

4.1 Targeted drug delivery (TDD)

TDD means the focused transportation of therapeutic agents into the diseased tissue without affecting other organs adversely. The enhanced permeability and retention (EPR) effect of the drug-conjugated nanocarrier system is a very crucial factor for TDD [65]. Ferrite NPs with their intrinsic responsiveness toward the external

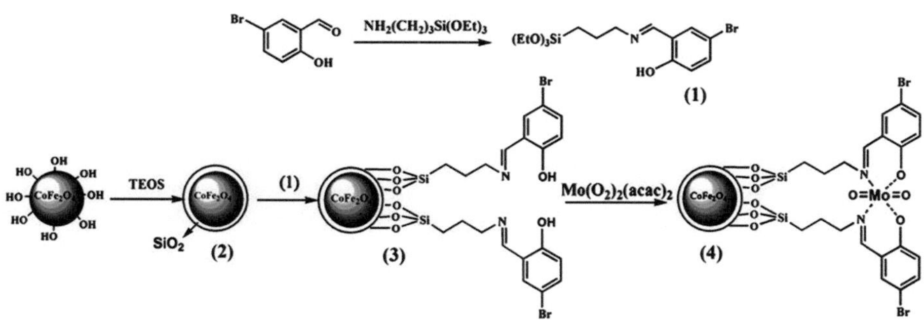

Figure 4.
Schematic illustration of the Mo-salenSi@Si-$CoFe_2O_4$ MNP preparation. Copyright: Catalysis letters [64].

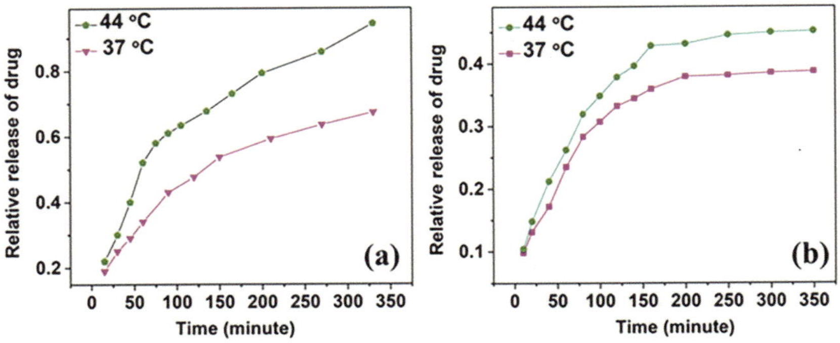

Figure 5.
Time-dependent DOX release profile of FA-coated cobalt ferrite MNPs at (a) pH = 5.5 and (b) pH = 7 as a function of time under temperatures 37 and 44°C in phosphate buffer solution. Copyright: Chemphyschem [8].

magnetic field smoothen the deposition, accumulation, and controlled release of immobilized drug to the targeted sites [66]. However, their surface modification accelerates this process by resisting aggregation, premature leakage of drug molecules, and removal of the MNPs by the organs *via* phagocytosis [65]. **Figure 5** shows the both pH- and temperature-responsive DOX release profile of the FA-coated CFMNPs. Here, FA-conjugation helped to retain the nanoparticles' stability even after the drug release experiment. Results showed that the relative drug release amount increased from ~38–60% by decreasing the solution pH from 7 to 5.5. Moreover, the heat generated by the applied AC magnetic field elevated the amount of released drug up to 95% [8].

FA modification significantly multiplies (~30 folds) the targeting effect of BSA-CFNP hybrids encapsulating bio-derived polyphenolic drugs hesperidin and eugenol. The magnetic CFNPs accounted for higher drug encapsulation efficiency (62.94% for hesperidin and 85.58% for eugenol). The BSA hybridization offered synergistic pH and magnetic responsiveness to the synthesized nanocarriers and helped the controlled and targeted release of drug molecules [18]. PEG-anchored CF-DA nanoparticles exhibited better cellular uptake into the cancerous cell line A549 than the unanchored ones. Results from the cell apoptosis, ROS generation, and actin cytoskeleton disruption study showed that CF-DA-PEG is able to produce more free radicals and cause mitochondrial dysfunction and actin cytoskeleton destruction inside the A549 cells [28].

4.2 Bioimaging

The magnetic spinel structure of ferrite NPs helps to shorten the spin-spin relaxation time (T_2-weighted) of surrounding water protons. This improves the contrast of the magnetic resonance (MR) image while real-time imaging of biological functions. The main problem here is the pharmacokinetic properties (biocompatibility, circulation time in the bloodstream, targetability, etc.) of the contrast agents [67]. Many researchers have tried to tailor the surface of the ferrite NPs with suitable functionalities to overcome this situation. PE-HBPG-FA hybrid-encapsulated $MnFe_2O_4$ MNPs are effective T2-weighted MRI contrast agents. This hyperbranched lipopolymer rendered water solubility and biocompatibility, whereas FA introduced tumor cell targetability to the synthesized spinel type $MnFe_2O_4$@PE-HBPG-FA MNPs. This composite exhibited a higher transverse relaxivity value (140.56 $mM^{-1}\ s^{-1}$) than conventional superparamagnetic iron oxides [17]. Disodium tartrate dihydrate (T)-functionalized and variable Gd^{3+}-doped $MnFe_2O_4$ (T-$MnGd_xFe_{2-x}O_4$) is an active fluorophore and possesses magnetic field-dependent photoluminescence (PL) properties. The ligand-to-metal charge transfer (LMCT) between the small organic ligand T and the dopant Gd^{3+} causes the high saturation magnetization of $TMnGd_{0.10}Fe_{1.90}O_4$ and a maximum PL intensity at about 417 nm. Moreover, this magneto-fluorophore showcases minimum cytotoxic effects and first-order degradation kinetics against bilirubin (BR). BR is a yellowish pigment responsible for hyperbilirubinemia or jaundice [68].

4.3 Biosensor

Functionalized ferrites are used as biosensing materials either by their direct application onto the transducer materials or by dispersion in the targeted environment [69]. For example, CH-modified n-$NiFe_2O_4$ comprises an efficient cholesterol biosensor in human serum samples. Basically, the biocompatible CH matrix helps in the homogeneous dispersion of n-$NiFe_2O_4$ NPs over the ITO bioelectrode and thus provides an ideal platform for cholesterol esterase (ChEt) and cholesterol oxidase (ChOx) immobilization. Electrochemical response studies prove the faster electron communication, high selectivity, and repeatability features of ChEt–ChOx/n-$NiFe_2O_4$ – CH/ITO bioelectrode [59]. Functionalization by a mixture of amine and thiol groups serves as an intermediate layer between the superparamagnetic cobalt ferrite nanoparticle core and gold nanoparticles' shell. The investigations by Marcos *et al.* show that the $M_2(4NH_2\text{-}SH)Au$ particles are able to immobilize single-stranded peptide nucleic acid (ssPNA) oligomers for the detection of single nucleotide polymorphisms (SNPs) in DNA [70].

4.4 Hyperthermia for cancer treatment

Hyperthermia treatment (also called thermal ablation or thermotherapy) is a process for treating cancerous cells locally under elevated temperatures (41–45°C). Ferrite NPs are able to produce heat of this temperature range while kept under an external alternating current (AC) magnetic field due to hysteresis loss and eddy current [8]. PEG-modified cobalt ferrite/hydroxyapatite immobilizing 5-Fluorouracil (FU) drug is a multimodel nanocarrier for synergistic chemotherapeutic and hyperthermia treatment. The presence of an AC magnetic field causes the heating of the magnetic core of this nanoparticle, which in turn compels the PEG matrix to release the immobilized drug molecules into the targeted sites [71]. Prashant *et al.* fabricated OA-coated $CoFe_2O_4$ NPs for magnetic fluid hyperthermia treatment. OA functionalization complements the great saturation magnetization and high permeability characteristics of $CoFe_2O_4$ NPs with its biocompatibility and high colloidal dispersive nature. OA-$CoFe_2O_4$nanocomposites exhibit reduced hypothermia temperature and magnetization values than the unmodified compounds in the water-ethylene glycol fluidic media [10]. Izabell *et al.* synthesized azelaic acid (AZA)-encapsulated manganese ferrite and zinc ferrite ($MnFe_2O_4$-AZA and $ZnFe_2O_4$-AZA) MNPs for high-performance hypothermic measurement. This hydrophilic outer layer of the MNPs imposes a direct effect on their high saturation magnetization values at low temperatures [72].

4.5 Tissue engineering

Considering the effect of Zn and Si in healthy bone tissue development and osteoblastic gene expression, Ashkan *et al.* formulated a multifunctional core-shell zinc silicate-$ZnFe_2O_4$ composite for bone tissue regeneration after implantation. Cell compatibility study using osteoblast cell line (MG63) revealed the concentration-dependent cell viability of the compound. Zn^{2+} ions are responsible for the antimicrobial nature of the complex against both the Gram-positive and Gram-negative bacterial strains [61]. Glutaric acid-functionalized cobalt ferrite ($CoFe_2O_4$-GAPT) MNPs are efficient agents for the top-down characterization of phosphoproteins during advanced disease diagnosis. SDS-PAGE analysis of the MNPs incubated pig cardiac phosphoproteins, and comparing the results with other non-treated samples showcased that the GAPT ligands facilitate the specificity and quantification efficacy of the MNPs [73].

5. Environmental pollution management

5.1 Wastewater treatment

5.1.1 By heavy metal adsorption

Zhiqiang *et al.* decorated the rice bran-derived biochar with ethylenediamine-functionalized $MgFeAlO_4$ (RB@$MgFeAlO_4$-NH_2) MNPs for Ni^{2+} elimination from the wastewater. The amine functionalization enhanced the sorption affinity of RB@ $MgFeAlO_4$-NH_2 *via* complexation with the Ni^{2+} ions, whereas the biochar employed the ion exchange mechanism during the purification process. Ni^{2+}sorption percentage is maximum at pH ~ 6 when Ni^{2+} ions can be electrostatically attached to the surface amino groups without the risk of precipitating from the solution [1]. **Figure 6** depicts the fabrication of La^{3+}-grafted and hexamethylendiamine-functionalized copper ferrite ($CuFe_2O_4$-2N-La) MNPs, which also showed high phosphate adsorption capacity (32.59 mg/g). A comparative study of the Langmuir adsorption isotherms of $CuFe_2O_4$ 2N and $CuFe_2O_4$-2N-La confirmed that La^{3+} ions have exerted a direct effect on increasing the phosphate decontamination rates from water bodies via the ligand coordination mechanism [7].

Figure 7 shows the reductive-adsorption and desorption behaviors of $MnFe_2O_4$ MNPs. Bilayered OP-optimized $MnFe_2O_4$ MNPs exhibit a highly specific sorption ability for uranium. Here, the unsaturated-unsaturated oleyl carbon chains form a compact layer around the $MnFe_2O_4$ particle surface, rendering superior colloidal stability of the MNPs in aqueous medium. This ordered organic coating of OP utilized both the chemisorption and reduction of U^{6+} into U^{4+} to maximize the sorption capacity of the MNPs [53].

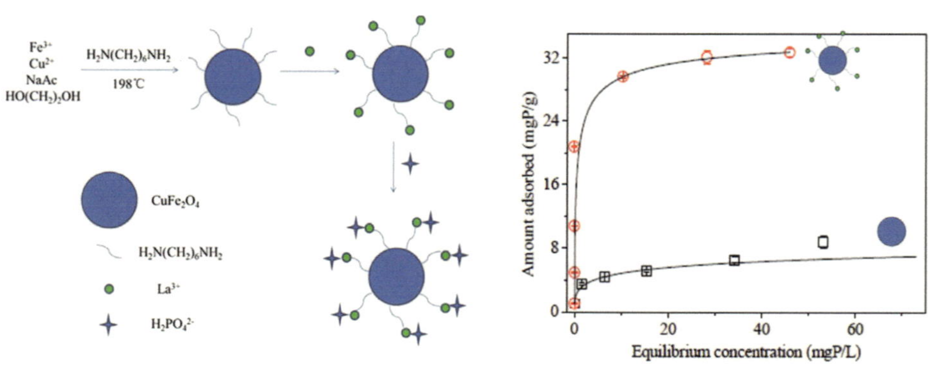

Figure 6.
Synthesis of $CuFe_2O_4$-2 N-La and comparison of phosphate adsorption efficiency of $CuFe_2O_4$2N and $CuFe_2O_4$-2 N-La using the Langmuir adsorption isotherms. Copyright: Science of the Total environment [7].

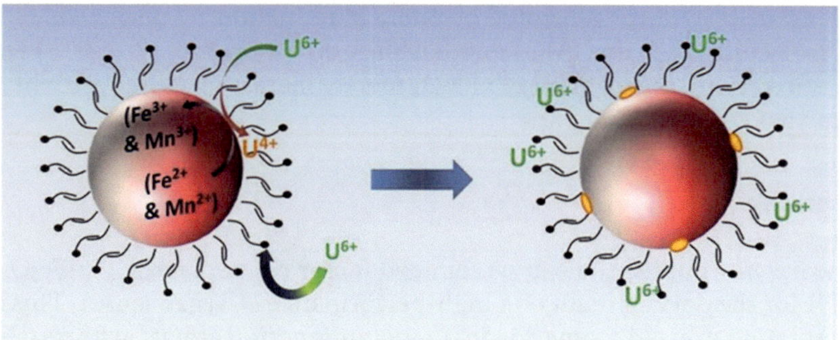

Figure 7.
Illustration of uranium adsorption by the reduction of U^{6+} into insoluble U^{4+} (orange) on the surface of the nanocrystalline $MnFe_2O_4$. Copyright: Journal of Materials Chemistry A [53].

5.1.2 Photocatalytic activity

Tudisco *et al.* anchored the visible light-sensitive TCPP molecules onto the surfaces of ferroelectric bismuth ferrite (BFO) MNPs. The resulting organic–inorganic composite BFO@TCPP is an efficient agent for the photocatalysis of organic dyes found in industrial wastewaters. TCPP helps enhance the surface-modified MNPs' catalytic activity by decreasing its band gap than free BFO and inhibiting raid recombination of the photogenerated electron–hole pairs. Moreover, the TCPP layer remained intact even after four degradation cyclic runs, indicating the stability and reusability of the BFO@TCPP particles [74]. TiO_2 is arguably one of the best semiconducting photocatalysts due to its chemical stability, nontoxicity, and enhanced photoreactivity. But it suffers from poor separation ability from the treated water. Wuyou *et al.* resolved this matter by functionalizing ferromagnetic $CoFe_2O_4$ nanoparticles with TiO_2 nanocrystals in a core–shell structure. The resulting nanocomposite's photocatalytic activity can be increased by increasing the percentage of TiO_2 in $TiO_2/CoFe_2O_4$ [62]. GO-modified copper ferrite ($GO/CuFe_2O_4$) nanocomposite is another worthy candidate for water remediation. The electron and hole pairs in a

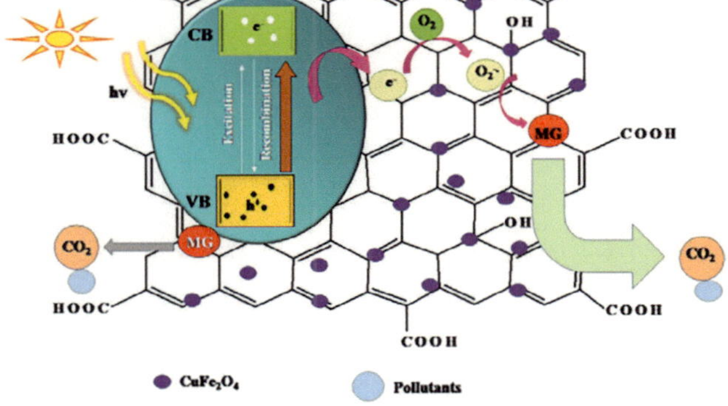

Figure 8.
Photocatalytic degradation mechanism of MG dye onto $GO/CuFe_2O_4$. Copyright: Materials today: Proceedings [75].

two-dimensional GO sheet generated by photo illumination are responsible for the superior malachite green dye (MG) degradability of GO/$CuFe_2O_4$ (62.37%) from the polluted water. Here, the magnetic $CuFe_2O_4$ merely increased the recyclability of this photocatalyst (**Figure 8**) [75].

5.2 Gas sensing

Xingwei *et al.* coupled the n-type semiconductor copper ferrite ($CuFe_2O_4$) with PANI for the development of a high-performance NH_3 gas sensor. This binary nanoformulation exerted a synergistic p-n heterojunction effect by decreasing the depletion layer, thus improving the response value and recovery time for NH_3 detection (**Figure 9**). The NH_3-detecting alarm device constructed based on this nanocomposite can selectively identify NH_3 gas of concentration as low as 5 ppm [37]. Another p-n heterojunction conducting material was synthesized by the *in situ* polymerization of PANI on NF nanoparticles for liquefied petroleum gas (LPG) detection. PANI modification mitigates the high-power consumption problem of NF. PANI-NF nanocomposite can operate at room temperature and retain its stability for over a month [38].

5.3 Supercapacitor (energy storage)

Barkha *et al.* synthesized glycol-functionalized reduced graphene oxide-cobalt ferrite (CoF-rGO)-based electrodes for superior energy storage application. $CoFe_2O_4$ possesses high structural anisotropy and specific capacitance (C_s) necessary for supercapacitor buildup but lacks in having low electrical conductivity. The glycol molecules improved the capacitance of CoF-rGO by increasing the surface wettability of the composite. Additionally, the incorporation of CoF into the graphitized structure facilitates the interlayer migration of ions by generating electric double layer during the intercalation-deintercalation process. As a result, rGO and CoF synergistically helped to retain 98% capacitance of the synthesized electrode material even after 2000 charge-discharge circles [76].

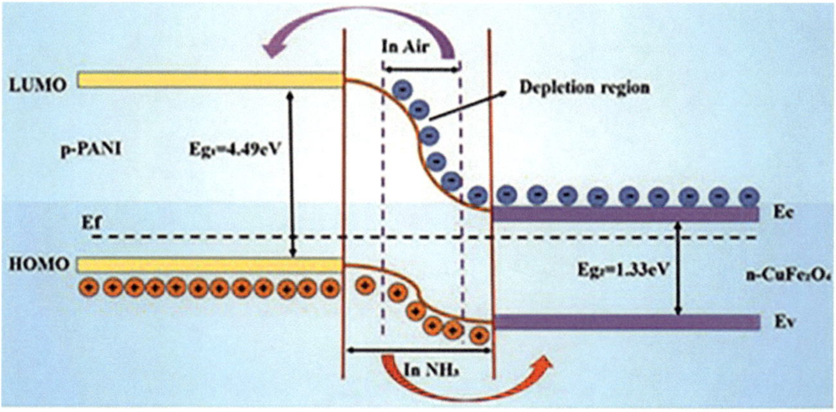

Figure 9.
Schematic illustration of the p-n heterojunction between PANI and $CuFe_2O_4$. Copyright: Sensors and actuators: B. Chemical [37].

Similarly, $NiFe_2O_4$ MNPs suffer from low power performance despite being anode materials in pseudocapacitor devices. For this reason, Neha *et al.* utilized a covalent functionalization process using aryl diazonium salt to restore the interfacial stability of the MNPs. The modified $NiFe_2O_4$ MNPs showcase strong metal–ligand bonds that eventually assist their overall magnetic and electrochemical behavior. Electrochemical studies showed that the diazonium-functionalized pseudocapacitor electrode exhibits substantially higher specific capacitance (~1279 Fg^{-1}) than bare $NiFe_2O_4$ and (~82–90%) after 2000 cycles [77].

6. Conclusion and perspectives

For successful therapeutic and environmental application of surface-functionalized ferrite nanoparticles, a clear understanding of the interaction between the functionalizing materials and the MNPs is crucial. Functionalization of the MNP surface effectively enhances the homogeneous dispersion, pH and temperature responsiveness, cellular uptake ability, and biocompatibility of ferrites for biomedical applications. However, for environmental protection management, introducing a new component onto the ferrite NPs surface would increase their sorption affinity, catalytic activity, recyclability, and specific capacitance according to the properties of the incoming materials. Therefore, it is very reasonable to think that functionalized ferrites hold great potentials. However, researchers need to be focused on fabricating multifunctional composites for synergistic drug delivery, tissue regeneration, and real-time bioimaging. When it comes to the practical applications using different biosafe compounds for different purposes that will significantly decrease the administration frequency, immunotoxicity and genotoxiciti, and the overall cost of the treatment. Similarly, for environmental protection, the designed models should be one-pot products, batch-to-batch reproducible, and easily quality controllable. Otherwise, industrialists and related policymakers would not be encouraged to adopt the green technology offered by these functionalized ferrite compounds. Finally, with the fast-paced experimental growth, it is time to concentrate on the commercialization of these materials. Several *in vivo* preclinical as well as clinical adaptations for the systematic validation of MNPs are currently in motion, and hopefully, we will witness the results very soon.

Author details

Tonmoye Sarkar Shathi and Abdur Rahman*
Research Laboratory of Polymer Colloids and Nanomaterials, Department of Chemistry, Faculty of Science, University of Rajshahi, Rajshahi, Bangladesh

*Address all correspondence to: rahmanabdur11@yahoo.com

IntechOpen

© 2023 The Author(s). Licensee IntechOpen. This chapter is distributed under the terms of the Creative Commons Attribution License (http://creativecommons.org/licenses/by/3.0), which permits unrestricted use, distribution, and reproduction in any medium, provided the original work is properly cited.

References

[1] Guo Z, Chen R, Yang R, Yang F, Chen J, Li Y, et al. Synthesis of amino-functionalized biochar/spinel ferrite magnetic composites for low-cost and efficient elimination of Ni(II) from wastewater. Science of the Total Environment. 2020;**722**:137822. DOI: 10.1016/j.scitotenv.2020.137822

[2] Maaz K, Karim S, Mashiatullah A, Liu J, Hou MD, Sun YM, et al. Structural analysis of nickel doped cobalt ferrite nanoparticles prepared by Coprecipitation route. Physica B Condens Matter. 2009;**404**:3947-3951. DOI: 10.1016/j.physb.2009.07.134

[3] Tomar D, Jeevanandam P. Synthesis of cobalt ferrite nanoparticles with different morphologies via thermal decomposition approach and studies on their magnetic properties. Journal of Alloys and Compounds. 2020;**843**:155815. DOI: 10.1016/j.jallcom.2020.155815

[4] Košak A, Makovec D, Žnidaršič A, Drofenik M. Preparation of MnZn-ferrite with microemulsion technique. Journal of the European Ceramic Society. 2004;**24**:959-962. DOI: 10.1016/S0955-2219(03)00524-7

[5] Goh SC, Chia CH, Zakaria S, Yusoff M, Haw CY, Ahmadi S, et al. Hydrothermal preparation of high saturation magnetization and Coercivity cobalt ferrite nanocrystals without subsequent calcination. Materials Chemistry and Physics. 2010;**120**:31-35. DOI: 10.1016/j.matchemphys.2009.10.016

[6] Chen D-H, He X-R. Synthesis of nickel ferrite nanoparticles by sol-gel method. Materials Research Bulletin. 2001;**36**:1369-1377. DOI: 10.1016/S0025-5408(01)00620-1

[7] Gu W, Li X, Xing M, Fang W, Wu D. Removal of phosphate from water by amine-functionalized copper ferrite chelated with La(III). Science of the Total Environment. 2018;**619-620**:42-48. DOI: 10.1016/j.scitotenv.2017.11.098

[8] Dey C, Ghosh A, Ahir M, Ghosh A, Goswami MM. Improvement of anticancer drug release by cobalt ferrite magnetic nanoparticles through combined PH and temperature responsive technique. ChemPhysChem. 2018;**19**:2872-2878. DOI: 10.1002/cphc.201800535

[9] Saharan P, Chaudhary GR, Mehta SK, Umar A. Removal of water contaminants by iron oxide nanomaterials. Journal of Nanoscience and Nanotechnology. 2014;**14**:627-643. DOI: 10.1166/jnn.2014.9053

[10] Kharat PB, Somvanshi SB, Khirade PP, Jadhav KM. Induction heating analysis of surface-functionalized nanoscale CoFe2O4 for magnetic fluid hyperthermia toward noninvasive cancer treatment. ACS Omega. 2020;**5**:23378-23384. DOI: 10.1021/acsomega.0c03332

[11] Thiruppathi R, Mishra S, Ganapathy M, Padmanabhan P, Gulyás B. Nanoparticle functionalization and its potentials for molecular imaging. Advanced Science. 2017;**4**:1600279. DOI: 10.1002/advs.201600279

[12] Ahmad F, Salem-Bekhit MM, Khan F, Alshehri S, Khan A, Ghoneim MM, et al. Unique properties of surface-functionalized nanoparticles for bio-application: Functionalization mechanisms and importance in application. Nanomaterials. 2022;**12**:1333. DOI: 10.3390/nano12081333

[13] Bohara RA, Thorat ND, Yadav HM, Pawar SH. One-step synthesis of uniform and biocompatible amine functionalized cobalt ferrite nanoparticles: A potential carrier for biomedical applications. New Journal of Chemistry. 2014;**38**:2979. DOI: 10.1039/c4nj00344f

[14] Irfan M, Zaheer F, Hussain H, Naz MY, Shukrullah S, Legutko S, et al. Kinetics and adsorption isotherms of amine-functionalized magnesium ferrite produced using sol-gel method for treatment of heavy metals in wastewater. Materials. 2022;**15**:4009. DOI: 10.3390/ma15114009

[15] Viltužnik B, Lobnik A, Košak A. The removal of Hg(II) ions from aqueous solutions by using thiol-functionalized cobalt ferrite magnetic nanoparticles. Journal of Sol-Gel Science and Technology. 2015;**74**:199-207. DOI: 10.1007/s10971-014-3596-x

[16] Li K, Xie L, Hao Z, Xiao M. Effective removal of Hg(II) ion from aqueous solutions by thiol functionalized cobalt ferrite magnetic mesoporous silica composite. Journal of Dispersion Science and Technology. 2020;**41**:503-509. DOI: 10.1080/01932691.2019.1591974

[17] Augustine R, Lee HR, Kim H, Zhang Y, Kim I. Hyperbranched Lipopolymer-folate-stabilized manganese ferrite nanoparticles for the water-soluble targeted MRI contrast agent. Reactive and Functional Polymers. 2019;**144**:104352. DOI: 10.1016/j.reactfunctpolym.2019.104352

[18] Uma Maheswari P, Muthappa R, Bindhya KP, Begum MS, K.M. Evaluation of folic acid functionalized BSA-CaFe2O4 Nanohybrid carrier for the controlled delivery of natural cytotoxic drugs hesperidin and eugenol. Journal of Drug Delivery Science and Technology. 2021;**61**:102105. DOI: 10.1016/j.jddst.2020.102105

[19] Morais PC, Santos RL, Pimenta ACM, Azevedo RB, Lima ECD. Preparation and characterization of ultra-stable biocompatible magnetic fluids using citrate-coated cobalt ferrite nanoparticles. Thin Solid Films. 2006;**515**:266-270. DOI: 10.1016/j.tsf.2005.12.079

[20] Kückelhaus S, Garcia VAP, Lacava LM, Azevedo RB, Lacava ZGM, Lima ECD, et al. Biological investigation of a citrate-coated cobalt–ferrite-based magnetic fluid. Journal of Applied Physics. 2003;**93**:6707-6708. DOI: 10.1063/1.1558665

[21] Fagundes DA, Leonel LV, Fernandez-Outon LE, Ardisson JD, dos Santos RG. Radiosensitizing effects of citrate-coated cobalt and nickel ferrite nanoparticle4.2.s on breast cancer cells. Nanomedicine. 2020;**15**:2823-2836. DOI: 10.2217/nnm-2020-0313

[22] Somvanshi SB, Kharat PB, Khedkar MV, Jadhav KM. Hydrophobic to hydrophilic surface transformation of Nano-scale zinc ferrite via oleic acid coating: Magnetic hyperthermia study towards biomedical applications. Ceramics International. 2020;**46**:7642-7653. DOI: 10.1016/j.ceramint.2019.11.265

[23] Nam PH, Lu LT, Linh PH, Manh DH, Thanh Tam LT, Phuc NX, et al. Polymer-coated cobalt ferrite nanoparticles: Synthesis, characterization, and toxicity for hyperthermia applications. New Journal of Chemistry. 2018;**42**:14530-14541. DOI: 10.1039/C8NJ01701H

[24] Khosravi HB, Rahimi R, Rabbani M, Maleki A, Mollahosseini A. Design, facile synthesis and characterization of porphyrin-zirconium-ferrite@SiO2 Core-Shell and catalytic application

in cyclohexane oxidation. Silicon. 2021;**13**:451-465. DOI: 10.1007/s12633-020-00454-w

[25] Kharazi P, Rahimi R, Rabbani M. Study on porphyrin/ZnFe2O4@polythiophene nanocomposite as a novel adsorbent and visible light driven Photocatalyst for the removal of methylene blue and methyl Orange. Materials Research Bulletin. 2018;**103**:133-141. DOI: 10.1016/j.materresbull.2018.03.031

[26] Grancharov SG, Zeng H, Sun S, Wang SX, O'Brien S, Murray CB, et al. Bio-functionalization of monodisperse magnetic nanoparticles and their use as biomolecular labels in a magnetic tunnel junction based sensor. The Journal of Physical Chemistry. B. 2005;**109**:13030-13035. DOI: 10.1021/jp051098c

[27] Jung K-W, Lee S, Lee YJ. Synthesis of novel magnesium ferrite (MgFe2O4)/biochar magnetic composites and its adsorption behavior for phosphate in aqueous solutions. Bioresource Technology. 2017;**245**:751-759. DOI: 10.1016/j.biortech.2017.09.035

[28] De D, Upadhyay P, Das A, Ghosh A, Adhikary A, Goswami MM. Studies on cancer cell death through delivery of dopamine as anti-cancer drug by a newly functionalized cobalt ferrite Nano-carrier. Colloids and Surfaces A: Physicochemical and Engineering Aspects. 2021;**627**:127202. DOI: 10.1016/j.colsurfa.2021.127202

[29] Wang G, Ma Y, Mu J, Zhang Z, Zhang X, Zhang L, et al. Monodisperse Polyvinylpyrrolidone-coated CoFe2O4 nanoparticles: Synthesis, characterization and cytotoxicity study. Applied Surface Science. 2016;**365**:114-119. DOI: 10.1016/j.apsusc.2016.01.031

[30] Kareem SH, Naji AM, Taqi ZJ, Jabir MS. Polyvinylpyrrolidone loaded-MnZnFe2O4 magnetic nanocomposites induce apoptosis in cancer cells through mitochondrial damage and P53 pathway. Journal of Inorganic and Organometallic Polymers and Materials. 2020;**30**:5009-5023. DOI: 10.1007/s10904-020-01651-1

[31] Jaberolansar E, Kameli P, Ahmadvand H, Salamati H. Synthesis and characterization of PVP-coated Co Zn Fe O 0.3 0.7 2 4 ferrite nanoparticles. Journal of Magnetism and Magnetic Materials. 2016;**404**:21-28. DOI: 10.1016/j.jmmm.2015.12.012

[32] Yardımcı FS, Şenel M, Baykal A. Amperometric hydrogen peroxide biosensor based on cobalt ferrite–chitosan nanocomposite. Materials Science and Engineering: C. 2012;**32**:269-275. DOI: 10.1016/j.msec.2011.10.028

[33] Bilas R, Sriram K, Maheswari PU, Sheriffa Begum KMM. Highly biocompatible chitosan with super paramagnetic calcium ferrite (CaFe2O4) nanoparticle for the release of ampicillin. International Journal of Biological Macromolecules. 2017;**97**:513-525. DOI: 10.1016/j.ijbiomac.2017.01.036

[34] Gingasu D, Mindru I, Patron L, Ianculescu A, Vasile E, Marinescu G, et al. Synthesis and characterization of chitosan-coated cobalt ferrite nanoparticles and their antimicrobial activity. Journal of Inorganic and Organometallic Polymers and Materials. 2018;**28**:1932-1941. DOI: 10.1007/s10904-018-0870-3

[35] Yavuz Ö, Ram MK, Aldissi M, Poddar P, Hariharan S. Synthesis and the physical properties of MnZn ferrite and NiMnZn ferrite–polyaniline nanocomposite particles. Journal of Materials Chemistry. 2005;**15**:810-817. DOI: 10.1039/B408165J

[36] Sadeghinia M, Shayeh JS, Fatemi F, Rahmandoust M, Ehsani A, Rezaei M. Electrochemical study of perlite-barium ferrite/conductive polymer Nano composite for super capacitor applications. International Journal of Hydrogen Energy. 2019;**44**:28088-28095. DOI: 10.1016/j.ijhydene.2019.09.085

[37] Wang X, Gong L, Zhang D, Fan X, Jin Y, Guo L. Room temperature ammonia gas sensor based on polyaniline/copper ferrite binary nanocomposites. Sensors and Actuators B: Chemical. 2020;**322**:128615. DOI: 10.1016/j.snb.2020.128615

[38] Kotresh S, Ravikiran YT, Vijayakumari SC, Thomas S. Interfacial P-n heterojunction of polyaniline-nickel ferrite nanocomposite as room temperature liquefied petroleum gas sensor. Composite Interfaces. 2017;**24**:549-561. DOI: 10.1080/09276440.2017.1241523

[39] Wang Y, Huang Y, Wang Q, He Q, Chen L. Preparation and electromagnetic properties of polyaniline(Polypyrrole)-BaFe12O19/Ni0.8Zn0.2Fe2O4 ferrite nanocomposites. Applied Surface Science. 2012;**259**:486-493. DOI: 10.1016/j.apsusc.2012.07.072

[40] Sagayaraj R, Aravazhi S, Chandrasekaran G. Effect of zinc content on structural, functional, morphological, resonance, thermal and magnetic properties of Co1−xZnxFe2O4/PVP nanocomposites. Journal of Inorganic and Organometallic Polymers and Materials. 2019;**29**:2252-2261. DOI: 10.1007/s10904-019-01183-3

[41] Sundararajan M, Sailaja V, John Kennedy L, Judith Vijaya J. Photocatalytic degradation of Rhodamine B under visible light using nanostructured zinc doped cobalt ferrite: Kinetics and mechanism. Ceramics International. 2017;**43**:540-548. DOI: 10.1016/j.ceramint.2016.09.191

[42] Sharma R, Singhal S. Structural, magnetic and electrical properties of zinc doped nickel ferrite and their application in photo catalytic degradation of methylene blue. Physica B Condens Matter. 2013;**414**:83-90. DOI: 10.1016/j.physb.2013.01.015

[43] Gallo J, García I, Padro D, Arnáiz B, Penadés S. Water-soluble magnetic Glyconanoparticles based on metal-doped ferrites coated with gold: Synthesis and characterization. Journal of Materials Chemistry. 2010;**20**:10010. DOI: 10.1039/c0jm01756f

[44] Nonkumwong J, Pakawanit P, Wipatanawin A, Jantaratana P, Ananta S, Srisombat L. Synthesis and cytotoxicity study of magnesium ferrite-gold Core-Shell nanoparticles. Materials Science and Engineering: C. 2016;**61**:123-132. DOI: 10.1016/j.msec.2015.12.021

[45] Farhadi S, Siadatnasab F. Synthesis and structural characterization of magnetic cadmium sulfide–cobalt ferrite nanocomposite, and study of its activity for dyes degradation under ultrasound. Journal of Molecular Structure. 2016;**1123**:171-179. DOI: 10.1016/j.molstruc.2016.06.032

[46] Xiong P, Zhu J, Wang X. Cadmium sulfide-ferrite nanocomposite as a magnetically recyclable photocatalyst with enhanced visible-light-driven photocatalytic activity and Photostability. Industrial and Engineering Chemistry Research. 2013;**52**:17126-17133. DOI: 10.1021/ie402437k

[47] Sahoo B, Devi KSP, Dutta S, Maiti TK, Pramanik P, Dhara D. Biocompatible mesoporous silica-coated superparamagnetic manganese ferrite

nanoparticles for targeted drug delivery and MR imaging applications. Journal of Colloid and Interface Science. 2014;**431**:31-41. DOI: 10.1016/j.jcis.2014.06.003

[48] Wang G, Ma Y, Wei Z, Qi M. Development of multifunctional cobalt ferrite/graphene oxide nanocomposites for magnetic resonance imaging and controlled drug delivery. Chemical Engineering Journal. 2016;**289**:150-160. DOI: 10.1016/j.cej.2015.12.072

[49] Yang Y, Shi H, Wang Y, Shi B, Guo L, Wu D, et al. Graphene oxide/manganese ferrite Nanohybrids for magnetic resonance imaging, Photothermal therapy and drug delivery. Journal of Biomaterials Applications. 2016;**30**:810-822. DOI: 10.1177/0885328215601926

[50] OyetadeOA,NyamoriVO,MartincighBS, Jonnalagadda SB. Effectiveness of carbon nanotube–cobalt ferrite nanocomposites for the adsorption of Rhodamine B from aqueous solutions. RSC Advances. 2015;**5**:22724-22739. DOI: 10.1039/C4RA15446K

[51] Wu H, Liu G, Wang X, Zhang J, Chen Y, Shi J, et al. Solvothermal synthesis of cobalt ferrite nanoparticles loaded on multiwalled carbon nanotubes for magnetic resonance imaging and drug delivery. Acta Biomaterialia. 2011;**7**:3496-3504. DOI: 10.1016/j.actbio.2011.05.031

[52] Cao Y, Mohamed AM, Mousavi M, Akinay Y. Poly(pyrrole-Co-styrene sulfonate)-encapsulated MWCNT/Fe–Ni alloy/NiFe2O4 nanocomposites for microwave absorption. Materials Chemistry and Physics. 2021;**259**:124169. DOI: 10.1016/j.matchemphys.2020.124169

[53] Lee SS, Li W, Kim C, Cho M, Lafferty BJ, Fortner JD. Surface functionalized manganese ferrite nanocrystals for enhanced uranium sorption and separation in water. Journal of Materials Chemistry A. 2015;**3**:21930-21939. DOI: 10.1039/C5TA04406E

[54] Pateiro M, Gómez B, Munekata PES, Barba FJ, Putnik P, Kovačević DB, et al. Nanoencapsulation of promising bioactive compounds to improve their absorption, stability, functionality and the appearance of the final food products. Molecules. 2021;**26**:1547. DOI: 10.3390/molecules26061547

[55] Rana S, Gallo A, Srivastava RS, Misra RDK. On the suitability of Nanocrystalline ferrites as a magnetic carrier for drug delivery: Functionalization, conjugation and drug release kinetics. Acta Biomaterialia. 2007;**3**:233-242. DOI: 10.1016/j.actbio.2006.10.006

[56] Casillas-Popova SN, Bernad-Bernad MJ, Gracia-Mora J. Modeling of adsorption and release kinetics of methotrexate from Thermo/magnetic responsive CoFe2O4–BaTiO3, CoFe2O4–Bi4Ti3O12 and Fe3O4–BaTiO3 Core-Shell Magnetoelectric nanoparticles functionalized with PNIPAm. Journal of Drug Delivery Science and Technology. 2022;**68**:103121. DOI: 10.1016/j.jddst.2022.103121

[57] Mathew AP, Oksman K. Processing of Bionanocomposites: Solution casting. In: Handbook of Green Materials: 2 Bionanocomposites: Processing, Characterization and Properties. 2014. pp. 35-52. DOI: 10.1142/9789814566469_0018

[58] Glazkova E, Bakina O, Rodkevich N, Mosunov A, Evstigneev M, Evstigneev V, et al. Antibacterial properties of PMMA functionalized with CuFe2O4/Cu2O/CuO nanoparticles. Coatings.

2022;**12**:957. DOI: 10.3390/coatings12070957

[59] Singh J, Roychoudhury A, Srivastava M, Chaudhary V, Prasanna R, Lee DW, et al. Highly efficient Bienzyme functionalized biocompatible nanostructured nickel ferrite–chitosan nanocomposite platform for biomedical application. The Journal of Physical Chemistry C. 2013;**117**:8491-8502. DOI: 10.1021/jp312698g

[60] Xu J, Yang H, Fu W, Du K, Sui Y, Chen J, et al. Preparation and magnetic properties of magnetite nanoparticles by sol–gel method. Journal of Magnetism and Magnetic Materials. 2007;**309**:307-311. DOI: 10.1016/j.jmmm.2006.07.037

[61] Bigham A, Foroughi F, Motamedi M, Rafienia M. Multifunctional Nanoporous magnetic zinc silicate-ZnFe2O4 Core-Shell composite for bone tissue engineering applications. Ceramics International. 2018;**44**:11798-11806. DOI: 10.1016/j.ceramint.2018.03.264

[62] Fu W, Yang H, Li M, Li M, Yang N, Zou G. Anatase TiO2 Nanolayer coating on cobalt ferrite nanoparticles for magnetic Photocatalyst. Materials Letters. 2005;**59**:3530-3534. DOI: 10.1016/j.matlet.2005.06.071

[63] Han Y, Lu Z, Teng Z, Liang J, Guo Z, Wang D, et al. Unraveling the growth mechanism of silica particles in the Stöber method: In situ seeded growth model. Langmuir. 2017;**33**:5879-5890. DOI: 10.1021/acs.langmuir.7b01140

[64] Kooti M, Afshari M. Molybdenum Schiff Base complex covalently anchored to silica-coated cobalt ferrite nanoparticles as a novel heterogeneous catalyst for the oxidation of alkenes. Catalysis Letters. 2012;**142**:319-325. DOI: 10.1007/s10562-012-0770-z

[65] Dahiya MS, Tomer VK, Duhan S. Metal–ferrite nanocomposites for targeted drug delivery. In: Applications of Nanocomposite Materials in Drug Delivery. Woodhead Publishing; 2018. pp. 737-760. DOI: 0.1016/B978-0-12-813741-3.00032-7

[66] Yang HW, Hua MY, Liu HL, Huang CY, Wei KC. Potential of magnetic nanoparticles for targeted drug delivery. Nanotechnology, science and applications. 2012. pp. 73-86

[67] Muhamad Arshad J, Raza W, Amin N, Nadeem K, Imran Arshad M, Azhar Khan M. Synthesis and characterization of cobalt ferrites as MRI contrast agent. Materials Today Proceedings. 2021;**47**:S50-S54. DOI: 10.1016/j.matpr.2020.04.746

[68] Chakraborty I, Majumder D, Rakshit R, Alam M, Mukherjee S, Gorai A, et al. Magnetic field-dependent photoluminescence of tartrate-functionalized gadolinium-doped manganese ferrite nanoparticles: A potential therapeutic agent for hyperbilirubinemia treatment. ACS Applied Nano Materials. 2021;**4**:4379-4387. DOI: 10.1021/acsanm.0c03073

[69] Rocha-Santos TAP. Sensors and biosensors based on magnetic nanoparticles. TrAC Trends in Analytical Chemistry. 2014;**62**:28-36. DOI: 10.1016/j.trac.2014.06.016

[70] Pita M, Abad JM, Vaz-Dominguez C, Briones C, Mateo-Martí E, Martín-Gago JA, et al. Synthesis of cobalt ferrite core/metallic shell nanoparticles for the development of a specific PNA/DNA biosensor. Journal of Colloid and Interface Science. 2008;**321**:484-492. DOI: 10.1016/j.jcis.2008.02.010

[71] Sangeetha K, Ashok M, Girija EK. Development of multifunctional cobalt

ferrite/hydroxyapatite nanocomposites by microwave assisted wet precipitation method: A promising platform for synergistic chemo-hyperthermia therapy. Ceramics International. 2019;**45**:12860-12869. DOI: 10.1016/j.ceramint.2019.03.209

[72] Crăciunescu I, Palade P, Iacob N, Ispas GM, Stanciu AE, Kuncser V, et al. High-performance functionalized magnetic nanoparticles with tailored sizes and shapes for localized hyperthermia applications. The Journal of Physical Chemistry C. 2021;**125**:11132-11146. DOI: 10.1021/acs.jpcc.1c01053

[73] Chen B, Hwang L, Ochowicz W, Lin Z, Guardado-Alvarez TM, Cai W, et al. Coupling functionalized cobalt ferrite nanoparticle enrichment with online LC/MS/MS for top-down phosphoproteomics. Chemical Science. 2017;**8**:4306-4311. DOI: 10.1039/C6SC05435H

[74] Tudisco C, Pulvirenti L, Cool P, Condorelli GG. Porphyrin functionalized bismuth ferrite for enhanced solar light photocatalysis. Dalton Transactions. 2020;**49**:8652-8660. DOI: 10.1039/C9DT04514G

[75] Yadav P, Surolia PK, Vaya D. Synthesis and application of copper ferrite-graphene oxide nanocomposite photocatalyst for the degradation of malachite green. Materials Today Proceedings. 2021;**43**:2949-2953. DOI: 10.1016/j.matpr.2021.01.301

[76] Rani B, Sahu NK. Electrochemical properties of CoFe2O4 nanoparticles and its RGO composite for supercapacitor. Diamond and Related Materials. 2020;**108**:107978. DOI: 10.1016/j.diamond.2020.107978

[77] Singh N, Malik A, Nohwar S, Jana R, Mondal PC. Covalent surface modification of nickel ferrite nanoparticles for electrochemical supercapacitor performance. New Journal of Chemistry. 2023;**47**:5308-5315. DOI: 10.1039/D2NJ05566J

Chapter 4

Potential of Ferrite-Based Nanoparticles for Improved Cancer Therapy: Recent Progress and Challenges Ahead

Jnanranjan Panda, Bhabani Sankar Satapathy, Abhishek Mishra, Biswabhusan Biswal and Pralaya Kumar Sahoo

Abstract

Recently, ferrite nanomaterials have emerged as a potent entrant in the biomedical field, especially in diagnosing and treating cancer in various organs because of their low toxicity, favorable magnetic properties, and biocompatibility. The conventional chemotherapy used for cancer treatment suffers from a deep setback because of the associated severe side effects produced in cancer patients during the treatment, such as bone marrow depression, hair fall, kidney damage, heart problems, neurological disorders, and others. Hence, in this context, ferrite nanomaterials provide the benefits of targeted delivery of a cytotoxic anticancer drug(s) to the specific tumor site using an external magnetic field, causing minimal side effects on healthy tissues. Another important benefit of using these nanomaterials lies in their ability to enhance the water solubility of hydrophobic drugs in order to extend the prolonged circulation of drugs in the blood and also to prevent fast renal excretion. Therefore, in this chapter, we will address the recent status and progress of ferrite-based nanomaterials in cancer therapy and will also cover the major challenges that hinder their translation from research to clinic.

Keywords: ferrite nanomaterials, cancer, nanomedicine, magnetic targeting, progress and challenges

1. Introduction

The interdisciplinary approach of nanomaterials brings researchers from physics, chemistry, and biology to put equal effort into synthesizing, understanding, and proceeding for various applications, from electronics, health, and environment to energy and information technology [1–5]. In this context, advancement in nanotechnology boosts human society with improved attributes of nanotechnological products. Nanomaterials are ultrafine particles of nanometer dimension (at least one dimension within 1–100 nm) with altered physico-chemical properties than macromaterials [6]. The magnificent properties of these nanomaterials provide a better future for their use in different diverse fields when compared to their bulk counterpart of the same composition because of the modified surface-to-volume ratio and tunable surface features [7, 8]. Among these, magnetic nanostructured materials (MNMs) have appeared as a robust entrant in the field of material synthesis and engineering because of their stupendous applications in the biomedical field, data storage, electronic devices (microwave, radiofrequency, and optoelectronics devices) [9–11], environmental remediation (catalysts, hydrogen storage), and others [12, 13] as summarized in **Figure 1**.

Furthermore, in the biomedical field, especially in the diagnosis and treatment of cancer, MNMs have gained proficient attention because of their excellent magnetic property and biocompatibility [14, 15]. The biomedical applications of MNMs rely on their ferromagnetic or superparamagnetic nature. This is because of the fact that in using an external magnetic field, these MNMs can be used for targeted delivery and release of anticancer drugs at the specific disease site. MNMs can be used for hyperthermia therapy, and they can also be used as a magnetic contrast agent in MRI diagnostic. For effective *in vivo* applications, such MNMs should be stable in an aqueous medium at physiological pH (7.4) and should be biocompatible [16–18]. In general, MNMs intended for *in vivo* must be conjugated or coated with a biocompatible polymer-like dextran, polyethylene glycol (PEG), polyvinyl alcohol (PVA), Poly

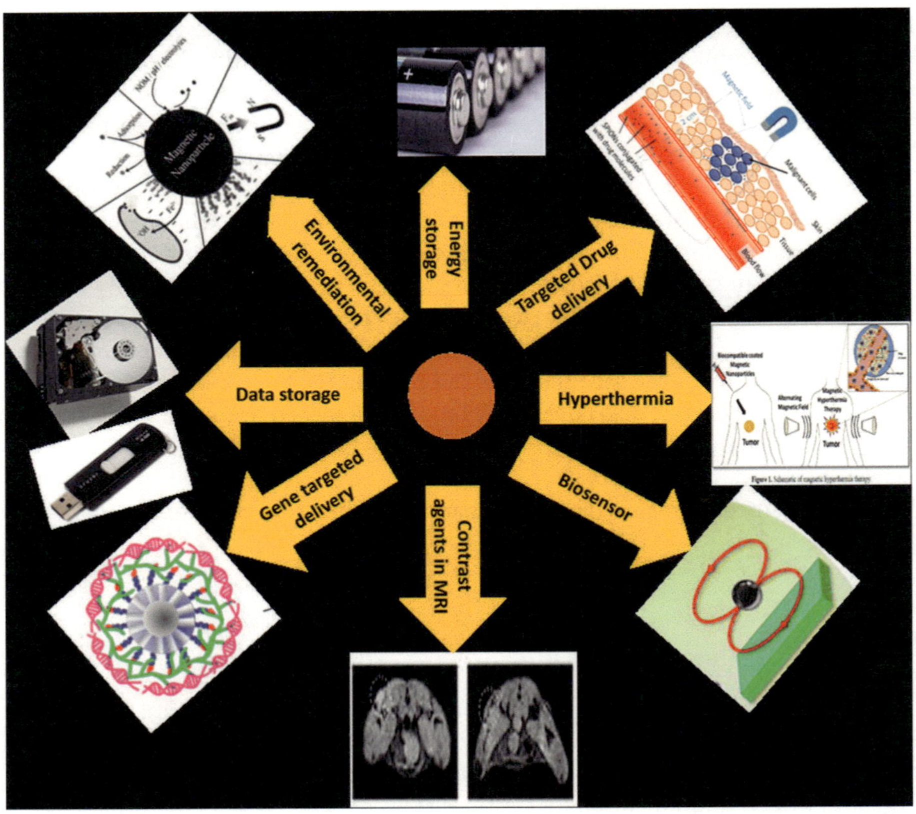

Figure 1.
Schematic representation of few technological applications of magnetic nanostructured materials.

D, L-lactic-co-glycolic acid (PLGA), and others during or after the synthesis process. Such type of surface-coated polymers on the MNMs helps to overcome the formation of large aggregates and also facilitates biodegradation when exposed to a biological system [19].

In a study from the International Agency for Research on Cancer (IARC), nearly 10 million cancer deaths were reported alone in 2020. Among all cancer cases, breast cancer tops the position with 684, 996 deaths in 2020 [20]. In recent times, similar to breast cancer, other types of cancer, such as lung, colon, and prostate cancer, have also been identified as one of the most fatal diseases with an alarming mortality rate across the globe. To date, the major prevention of cancer relies on early diagnosis, surgery, radiation, and drug therapy; however, once the cancer cells start metastasizing, drug therapy (chemotherapy) remains to be the only available option. However, chemotherapy seriously fails to meet the expected outcome in cancer patients, owing to its severe healthy tissue toxic effects. Many cancer patients even fail to complete chemotherapy regimens because of unbearable side effects, including immune suppression, risk of heart attack, hair fall, kidney problems, and others. It would not exaggerate to claim that the toxicity of conventional drug therapy is directly attributed to the off-target distribution of anticancer drugs to healthy tissues [21]. Hence, the main difficulties of cancer treatment lie in the cancer cell-specific delivery of cytotoxic drugs with diminished side effects. Consequently, in recent years, major efforts have been devoted to develop and design novel drug carriers that will overcome the downsides of conventional chemotherapy.

Many novel drug carriers, such as nanoliposomes [22], polymeric nanoparticles [23], carbon-based systems [24], gold nanoparticles [25], and others, have been reported over the past few years to enhance treatment outcomes in cancer patients. Despite the development of numerous types of nanocarrier for cancer therapy, magnetite (Fe_3O_4), manganese ferrite ($MnFe_2O_4$), and cobalt ferrite ($CoFe_2O_4$)-based MNMs because of their biocompatibility, favorable magnetic properties, higher chemical stability, ease of surface modification, and others are deliberated preferable over other nano drug carriers for tumor cell-specific delivery of cytotoxic anticancer drugs [19, 21]. MNMs provide the benefits of targeted delivery of anticancer drug(s) to the specific tumor site by applying an external magnetic field, causing minimal side effects on healthy tissues [19, 25]. Another important benefit of using such MNMs lies in their ability to enhance the water solubility of hydrophobic anticancer drugs, extend the prolonged circulation of drugs in the blood, and also prevent fast renal excretion [26, 27]. The nanoscale dimensions of the MNMs allow them not only to pass through the narrowest blood vessels but also to cross through cell membranes (paracellular transport) [21]. The present chapter deals with recent developments in the ferrite-based nano drug carriers investigated over the past years for cancer therapy. Along with some light has also been thrown on the challenges associated with MNMs for their clinical feasibility and large-scale technology transfer.

2. Anticancer drug delivery through magnetic nanocarriers/magnetic drug targeting

The drug-targeting concept for the cell-specific treatment of cancer has been a widely expanding field in medical science research. The idea of "magnetic drug delivery" by employing MNMs as drug carriers was proposed at the end of the 1970s by Widder and Senyei et al. [28, 29]. Despite the development of various nanocarriers, magnetic-mediated drug nanocarriers are becoming increasingly popular for the targeted delivery of cytotoxic anticancer drugs. Here, the therapeutic agents are encapsulated or attached to MNMs using a polymeric layer, which are then guided to the target site using an external magnetic field. The advantages of using these polymers include helping in the conjugation with the biological ligands for better tumor specificity and also enhancing the circulation time of MNMs in the blood (increased half-life) [19, 25]. A variety of anticancer drugs, such as cisplatin, methotrexate, danorubicin, doxorubicin, and paclitaxel, have been loaded into the organic/inorganic scaffold of surface-modified MNMs for drug delivery application. MNMs have the ability to carry a large dose of anticancer drug to attain high, local concentration, and avoid other inimical side effects resulting from high drug doses in other healthy tissues of the body. The low-cost factor and the simple synthesis technique further add value for their future large-scale production. After entering the body, MNMs can be directed to a specific solid tumor site under the influence of an external magnetic field. For use in the biomedical field, MNMs of homogeneous size and uniform shape are required [30, 31]. At the same time, good crystallinity and phase control of MNMs are also desired. MNMs must also have biocompatibility, good thermal stability, and suitable magnetic moment. MNMs can be coated with biomolecules or biodegradable polymers to increment the residence time in the blood circulation systems or to make them interact with a cell or a biological entity [19]. Drug-loaded MNMs are injected into laboratory animals (rats/mice) *via* an intravenous (IV) route. Around the tumor location, an external magnetic field is then applied where the maximum amount of the administered drug is supposed to be localized under the influence of the applied magnetic field. The drug from MNMs is then released by the enzymatic activity or changes in temperature/pH or other physiological conditions [32]. Furthermore, the smaller size of MNMs possesses

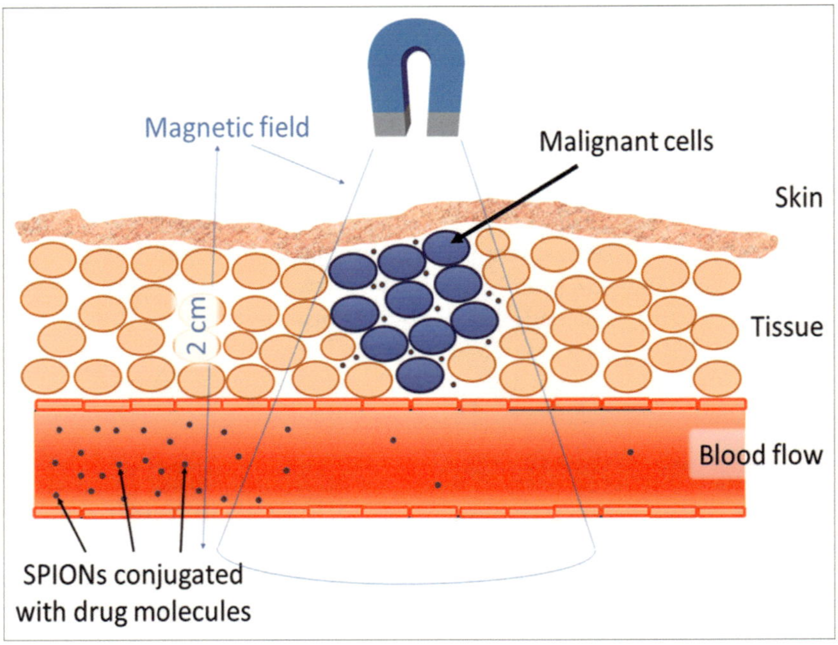

Figure 2.
Schematic representation of magnetic drug targeting under the influence of an external magnetic field.

many additional benefits, such as increased internalization across fenestrated tumor vessels, smart escape from RES uptake, improved blood circulation, and others [15]. The size and surface charge of MNMs have robust effects on the pharmacokinetics and the bioavailability of the loaded cargo within the body. The first phase clinical trial of magnetic targeting was performed in cancer patients using epirubicin-loaded MNMs [33]. The mechanism of drug targeting by ferrite-based nanocarriers at tumor locations has been represented in **Figure 2**.

3. Design of MNMs for biomedical applications

For drug delivery applications, the used MNMs consist of several including magnetic core, outer coating, and surface functional groups. A proper selection of magnetic core is essential, which defines the capabilities of the system, for example, heating; sensing, and others. The core portion of the MNMs should have a homogeneous size and uniform shape with a superparamagnetic or ferromagnetic nature [25, 34]. The superparamagnetic behavior ensures that the MNMs exhibit zero remanent magnetization when there is no applied field and a high magnetization under an externally applied magnetic field. Proper surface modification of MNMs is crucial to retain their colloidal stability and biocompatibility in any biological environment [35]. In selecting a magnetic core, a few key features such as particle size, functionality, stability, biocompatibility, and superparamagnetic nature are usually taken into consideration. The size distribution of the MNMs must be as small as possible (about 10–200 nm) having a monodisperse pattern. This is because all the physicochemical and magnetic properties strongly depend on the size and morphology of the MNMs. The hydrodynamic size of MNMs (which includes the total diameter of MNMs and the protective coating thickness) is also crucial for improved blood circulation time. Various kinds of inorganic materials, such as pure metal, metal alloys, iron oxide, and core-shell structures, have been explored for biomedical applications. Among MNMs, transition metals such as Fe, Co, Ni, and Mn are good options because of their high magnetization values, which is one of the key parameters for high performance hyperthermia and MRI applications. However, they are unstable and easily oxidized during synthesis time [7, 8]. These unavoidable problems make them unsuitable for biomedical applications. However, iron oxide-based magnetic nanoparticles such as magnetite (Fe_3O_4) [36, 37], cobalt ferrite [38, 39], manganese ferrite [40, 41], and zinc ferrite [42, 43] have received substantial attention for their promising application in biomedical fields. This is because of their biocompatibility and biodegradability along with their optimum magnetic property. Iron oxide nanoparticles (IONP) are stable and have good magnetization. It has an outstanding chemical stability, superb electrical insulation, moderate saturation magnetization, and a high Curie temperature. The similarity in magnetic behavior and other properties of both magnetite and maghemite make them difficult to distinguish. Fe_3O_4 has an inverse spinel structure where Fe^{+2} ions are occupied by octahedral sites, and Fe^{+3} ions are equally distributed between octahedral and tetrahedral sites and can be represented by $[Fe^{3+}]$ A $[Fe^{3+}, Fe^{2+}]BO_4$. In Fe_3O_4, there is an equal number of Fe^{+3} ions in the octahedral and tetrahedral sites, which compensate for each other, and the resulting magnetization arises only from the uncompensated Fe^{+2} ions in octahedral sites. However, maghemite (which has the same spinel structure as magnetite) forms as a result of the oxidation of magnetite and contains only Fe^{+2} ions distributed randomly over the octahedral and tetrahedral sites. The magnetization of maghemite arises from uncompensated Fe^{+3} ions [44, 45]. The cation distribution in the tetrahedral and octahedral sites strongly depends upon the methodology adopted for the synthesis. The other alternative ferrites, namely, cobalt ferrite, manganese ferrite, and zinc ferrite are also investigated for various types of biomedical applications. Here, Fe^{+2} ions are fully or partially altered by transition metals such as (M = Co, Mn, and Zn) in spinel structure, then they are represented by the general formula MFe_2O_4. Currently, more than 20,000 studies on IONPs have been reported, wherein the number of papers and biological effects of other MNMs, such as $CoFe_2O_4$, $MnFe_2O_4$, and $ZnFe_2O_4$, were less.

4. Synthesis of nanostructured magnetic materials

Over the past few years, considerable efforts have been devoted to the preparation of MNMs and understanding their properties for utilizing them in different areas of application. It is well-established that for a nanoparticle system, the physicochemical properties along with its biological fate strongly depend on the chemical composition and morphology of the particles, which are very sensitive to the synthesis route and surface engineering strategy of the given nanoparticle. Therefore, it is very crucial for controlling the synthesis conditions and surface engineering of MNMs, which governs their physicochemical properties, colloidal stability, and biocompatibility. Although numerous techniques have been reported for the synthesis of a wide variety of magnetic nanoparticles, some of the techniques are single-step, whereas others are multi-step techniques. All these techniques have both advantages and disadvantages, and none of these shows an authentic solution for all types of MNMs. Thus, one has to consider whether the chosen route is suitable for synthesizing the specific MNMs in a given environment using the available instrumental and synthesis facilities. Most of the synthesis techniques involve simple, basic inorganic chemistry methods. Instead of discussing all synthesis routes, here, we only discuss the synthetic procedure and its corresponding formation mechanism of co-precipitation and hydrothermal techniques. These chemical routes can be used to prepare steady and size-controlled MNMs when compared with other physical methods, such as gas-phase deposition.

4.1 Co-precipitation method

This method is one of the simplest techniques for synthesizing MNMs. The IONPs can be prepared using the method of precipitation from aqueous solution containing ferric and ferrous salts (in a 2:1 stoichiometric ratio) at a temperature (70–90°C) under an inert atmosphere by the addition of a base. The size and morphology of MNMs strongly depend on the type of salts, such as chlorides, sulphates, nitrates, and others, Fe^{2+} to Fe^{3+} concentration ratio, reaction temperature, and pH of the solution [46, 47]. The pH of the solution during formulation development usually remains between 9 to 14 [48, 49]. The chemical reaction for this process can be expressed as $Fe^{2+} + 2Fe^{3+} + 8OH^- = Fe_3O_4 + 4H_2O$.

Magnetite is not stable enough and easily converted into maghemite in the presence of oxygen. The chemical reaction is written as $Fe_3O_4 + 2H^+ = Fe_2O_3 + Fe^{2+} + H_2O$. The surfaces of MNMs are coated with organic polymers during the precipitation process to prevent the oxidation caused by air, as well as the formation of aggregates. MNMs with broad particle sizes having irregular morphology are prepared by this technique. However, the preparation of mixed oxide via this technique is less straightforward because the different metals precipitate at different pH values. In this synthesis technique, only kinetic factors control the growth of the crystal. However, getting a narrow size distribution with a homogenous shape is a challenging task using this synthesis route. In terms of simplicity, the co-precipitation method is considered a preferred route for IONP synthesis.

4.2 Hydrothermal method

Like the co-precipitation technique, hydrothermal method is relatively less explored for the synthesis of MNMs, although it allows the synthesis of high-quality particles. Polar solvents, such as water, methanol, or isopropanol as well as organic solvents, are used in this technique. The MNMs are formed by the dissolution and crystallization mechanism. In this technique, mixture consisting of $FeCl_3$, ethylene glycol, sodium acetate, and polyethylene glycol is stirred vigorously to form a clear solution, then sealed in a Teflon-lined stainless steel autoclave, and heated to and maintained at (130–250°C) temperature for 8 to 72 hours [8]. The precursor solution is poured into the Teflon chamber in such a way that 80% of it is filled. Finally, the temperature of the autoclave is allowed to cool down to room temperature and the resultant supernatant solution is washed to remove unused surfactants, impurities, and unreacted precursors. Here, ethylene glycol is used as a high-boiling point reducing agent, sodium acetate as an electrostatic stabilizer to prevent particle agglomeration, and polyethylene glycol as a surfactant against particle agglomeration [8]. The parameters, such as heating temperatures, reaction timings, and the ratio of the precursor to surface coatings, are tailored to obtain biocompatible MNMs with various sizes and shapes. In this way, different types of MNMs can be fabricated with tunable sizes with a high degree of crystallinity [50, 51]. Other metal oxide nanoparticles can be also prepared using this technique.

5. Stabilization of nanostructured magnetic materials

Stability is a pivotal requirement for almost any technological application of MNMs. Bare MNMs tend to agglomerate and form clusters in order to reduce their large-surface area-to-volume ratio. Therefore, the protective coating of MNMs is a mandatory requirement to achieve good colloidal stability with non-agglomerated particles. This is usually done by developing a core-shell structure, wherein the bare MNMs can be coated with a material forming a shell, hence, isolating the magnetic core from the outside environment. MNMs can be coated using an organic or inorganic material. For organic coating, organic materials including polymers, oleic acid, oleylamine, dodecyl amine, and others are used. For coating with inorganic components (viz. silica, carbon, and Au), and others are used. For polymer coating of MNMs, natural polymers, such as chitosan, starch, dextran, albumin, and others, and synthetic polymers such as PEG, PVA, PLGA, and others can be employed. To avoid agglomeration, the surfaces of MNMs are passivated by polymer coatings. Here, the polymers are adsorbed on the surface of MNMs by forming a single or double layer. This layer causes steric repulsion to balance the magnetic and the van der Waals forces acting on the particles. MNMs stabilized by polymers are not stable in air and can be easily separated by acidic solution, affecting their magnetization value. At higher temperature, polymer-coated MNMs exhibit low intrinsic stability, which can even be enhanced by a possible catalytic action of the metallic cores. Therefore, non-polymeric coated materials are sometimes a better choice instead of polymer-coated materials to provide better colloidal stability and prevent agglomeration of MNMs.

6. Recent research findings on MNMs in drug delivery system

The idea of "magnetic drug targeting" by employing MNMs as drug carriers to specific tumor sites was proposed by Widder and Senyi in late 1970 [28, 29]. Following their early investigations, the anticancer efficacy of this approach was proved in several minor animal tests against various types of malignancies, such as lung, breast, prostate, and brain cancer and even resulted in a limited number of animal studies [52]. Despite these efforts, however, this technique has yet to be standardized at the pre-clinical level. One of the primary reasons is that present MNMs have limited payload capacity because payload can either be affixed to the surface or incorporated in the double-layer covering that surrounds MNMs. Lubbe et al. conducted the first animal MNM investigation, in which tiny volumes of ferrofluid were utilized as delivery systems to concentrate epirubicin in tumors [33]. The study found that there were no significant anomalies caused by the ferrofluid. Therefore, it was determined that the magnetic fluid was safe and could be utilized to treat cancer. According to the findings, the same study team conducted Phase I and Phase II clinical studies between 1996 and 2001 [33, 53]. The outcomes showed that patients tolerated magnetic drug targeting with epirubicin well; and that in around 50% of patients, the nanoparticles were successfully targeted to the tumors. Methotrexate-conjugated IONPs were created by Kohler et al. [53] and have the potential to serve as drug carriers in controlled drug release as well as contrast-enhancing agents [54].

Superparamagnetic $CoFe_2O_4$ NPs coated with folic acid were created by Mohapatra et al. [54], and their anticancer effectiveness was tested on HeLa cells. They noticed that $CoFe_2O_4$ NPs did not significantly alter the proliferation at concentrations up to 20 mg/mL in comparison with the control, indicating the nontoxicity of $CoFe_2O_4$ NPs [55]. According to Georgiadou et al. [55], $CoFe_2O_4$ NPs with oleylamine dramatically lowered their toxicity in normal cells when compared to malignant cells [56].

RBCs and polymorphonuclear (PMN) leukocytes from human blood, as well as an 8-week-old Swiss male mice, did not exhibit any in vivo or in vitro toxicity when exposed to spherical $CoFe_2O_4$ NPs with a size of 20 to 100 nm, according to L.F. Cotica et al. [56]. No discernible toxicity was seen in studies of in vitro toxicity on RBCs and PMN leukocytes for periods of 24 and 6 hours at concentrations of 0.02, 0.01, 0.005, and 0.0025 mg/mL [57].

Because of the increased hyperthermia and coercivity created by mixing the magnetic component (Co) with Fe_3O_4 in Matsuda et al.'s [57] study, spermine-coated $CoFe_2O_4$ nanoparticles exhibit more potent anticancer activity against the MCF-7 cell line than the Fe_3O_4 NPs [38].

$CoFe_2O_4$ NPs were developed hydrothermally by Ansari et al. [38], and their anticancer activity against MCF-7 cells and healthy cells was studied. According to their research, $CoFe_2O_4$ nanoparticles were harmless for normal cells and only weakly anti-proliferative against MCF-7 cells [58]. The L-cysteine coating improves the colloid stability and biocompatibility of $CoFe_2O_4$ NPs according to Wang et al. [58]. They claimed that $CoFe_2O_4$ NPs coated with L-cysteine could transport doxorubicin at a concentration of 0.62 mg/mg in the form of a nanocarrier. They also demonstrated that medication release is almost twice as great under acidic circumstances as it is in neutral ones. Additionally, their findings showed that even at concentrations as high as 150 g/mL, HeLa cells only exhibited around 10% of the expected rate of apoptosis after 24 hours of incubation [59].

Similar to this, Fan et al. [59] suggested using silica-coated $CoFe_2O_4$NPs as a targeted nanocarrier for DOX delivery to the HeLa cell line. They demonstrated the pH dependence of the nanodrug delivery system's drug release performance and its rise with a lowering pH value [60]. Docetaxel-loaded PLGA-PEG-based $CoFe_2O_4$ nanodrug delivery system was created by Panda et al. [15] who also tested the in vitro cytotoxicity against the MCF-7 and MDA-MB-231 breast cancer cell lines. Drug-loaded NPs were reported to have respective IC_{50} values of 15.58 g/ml and 14.05 g/ml against MDA-MB 231 and MCF-7 cell lines [15]. To test the system's potential for magnetic drug targeting, Gaihre et al. [60] synthesized doxorubicin-loaded IONPs in gelatine [61]. In a research on drug release under various pH conditions, an acidic medium showed a persistent pattern of drug release compared with a neutral media.

In the transport of docetaxel to prostate cancer cells, Ling et al. [61] established PLGA-based IONPs functionalized by PEG [62]. Their research showed that the formulation had improved drug release characteristics and increased drug loading effectiveness (6.08%). They concluded by saying that these nano-drug formulations would be appealing for future nanomedicine development of multifunctional vesicles for simultaneous targeted imaging and drug delivery vehicles for prostate cancer treatment.

Because of the considerable cellular absorption of nanoparticles on CT-26 (Colon) cell line and their magnetic characteristics, Schleich et al. [62] demonstrated that paclitaxel-loaded PLGA-based Fe_3O_4 NPs might be employed as tumor-targeting MRI contrast agents [63]. Future nanomedicine applications might combine molecular imaging, drug delivery, and real-time treatment response monitoring because of the NPs' multifunctionality. Maghemite NPs with a silica coating were made by Rudzka et al. [63], and then they were functionalized with gold [64]. Using a magnetic drug delivery device, they saw a maximum loading of 80 mol/g for doxorubicin. The cytotoxicity investigation also showed that a medication delivery method based on maghemite is more effective against colon cancer than against liver cancer. By developing iron oxide-based PLGA NPs, Zhou et al. [64] hypothesized that these microspheres may be employed as contrast agents for dual imaging and to improve the effects of high-intensity focused ultrasound ablation on liver tissue [65].

Curcumin-loaded chitosan-modified Fe_3O_4 NPs were generated by Pham et al. [65], and their anticancer effectiveness against the A549 cell was examined. The IC_{50} value for the A549 cell was 73.03 g/ml. The findings demonstrated that the altered Fe_3O_4 NPs might be utilized as a nanodrug carrier for the treatment of cancer [66]. Cui et al. [66] generated magnetic PLGA NPs modified with transferrin receptor-binding peptides by co-encapsulating MNPs with a dual medicine (paclitaxel and curcumin). Comparing the dual-targeting effects with the non-targeting NPs, they found that cellular uptake studies showed a > 10-fold increase, and brain delivery showed a > 5-fold improvement [67].

A hydrophobic surface on a magnetic nanocarrier of Fe_3O_4 was created by Pourjavadi et al. to enable the adsorption of significant quantities of anticancer medicines [68]. To increase the colloidal stability and biocompatibility, the drug-loaded magnetic nanocarriers were coated with an alginate polymer shell. Additionally, they noted that alginate shells are removed from nanocarrier surfaces in acidic media, resulting in a higher rate of drug release than in neutral media wherein alginate shells are stable. To improve the delivery of docetaxel to breast cancer cells, Panda et al. (2018) created a PLGA-PEG-based Fe_3O_4 nanodrug delivery system [19]. Drug-loaded NPs were found to have an IC50 value of 18.4 g/ml. The proposed nanodrug formulation displayed acceptable cytotoxicity against MCF-7 cells and a predominate uptake by MCF-7 cells throughout a 0.5-hour incubation period. Additionally, the team investigated the *in vivo* pharmacokinetics of docetaxel-loaded Fe_3O_4 in comparison with the drug's free form. The cytotoxic efficiency of curcumin-loaded PLGA-Fe_3O_4 microspheres against HeLa (cervical cancer) cells was enhanced in comparison with the bare curcumin and magnetic microspheres according to Ayyanaar et al. [68]. They suggest using magnetically tailored medicine delivery devices using their magnetic nanocomposite [68].

Doxorubicin-loaded PLGA-Fe_3O_4 core-shell nanocomposite was created by Zhu et al. [69] and was employed as a dual drug delivery method and an MPI quantification tracer. Using the nanocomposite injection, observing the drug release, and evaluating the resulting tumor cell mortality, they also carried out *in vivo* drug release monitoring in a cancer treatment context using a mouse breast cancer model. Compared with existing monitoring techniques, this study offers a better method for *in vivo* drug release monitoring. A pancreatic cancer cell line and an orthotopic xenograft mouse model were used by Khan et al. (2019) to examine the effectiveness of the curcumin-loaded Fe_3O_4 NPs to overcome gemcitabine resistance and enhance its therapeutic potential [70]. Their findings showed that gemcitabine and curcumin-loaded Fe_3O_4 NPs had powerful synergistic effects in suppressing human pancreatic cancer cells as well as cancer stem cells. One significant advancement is the development of targeted drug delivery systems using ferrite-based nanoparticles. These nanoparticles can be functionalized with specific ligands or antibodies that selectively bind to cancer cells, allowing for targeted delivery of therapeutic agents directly to the tumor site. This approach minimizes off-target effects and improves the efficacy of treatment while reducing systemic toxicity [71]. Ferrite NPs can generate heat when exposed to an alternating magnetic field. This property has been utilized in hyperthermia therapy, wherein the nanoparticles are selectively delivered to the tumor site and then heated to induce localized tumor cell death. Recent studies have demonstrated enhanced therapeutic outcomes by using ferrite nanoparticles with improved heating capabilities and biocompatibility [72]. Recent progress has focused on developing ferrite-based nanoparticles with enhanced imaging capabilities, such as improved signal intensity and prolonged circulation time. These advancements allow for better visualization of tumors and monitoring of treatment response [73].

7. Challenges of magnetic drug targeting

- **Biocompatibility and toxicity:** Ferrite nanoparticles need to be biocompatible to ensure minimal adverse effects on healthy tissues. Their potential toxicity, particularly when used at high concentrations or for long durations, requires careful evaluation and optimization.

- **Pharmacokinetics and biodistribution:** Understanding the behavior of ferrite nanoparticles *in vivo*, including their pharmacokinetics, biodistribution, and clearance pathways, is crucial for safe and effective therapeutic applications. Addressing concerns related to long-term stability and potential accumulation in organs is essential.

- **Scalability and manufacturing:** Developing scalable synthesis methods for ferrite nanoparticles with consistent size, shape, and surface properties is essential for clinical translation. The cost-effectiveness of large-scale production should also be considered.

- **Clinical translation:** Although preclinical studies have demonstrated promising results, clinical trials are necessary to assess the safety, efficacy, and long-term outcomes of ferrite-based nanoparticle therapies. Regulatory and approval processes need to be navigated to bring these therapies to the clinic.

- **Combination therapies:** Integrating ferrite-based nanoparticles with other treatment modalities, such as chemotherapy, immunotherapy, or radiation therapy, requires careful optimization and understanding of synergistic effects, dosing regimens, and potential drug interactions.

- **Tumor heterogeneity:** Cancer is a highly heterogeneous disease, and the efficacy of ferrite-based nanoparticle therapies may vary across different tumor types and subtypes. Tailoring treatments to individual patient characteristics and addressing interpatient variability pose significant challenges.

- **Stability and controlled release:** Ensuring the stability of ferrite nanoparticles during storage and transportation, as well as achieving controlled release of therapeutic payloads, remains a technical challenge that needs to be overcome.

Addressing these challenges will pave the way for the successful implementation of ferrite-based nanoparticles in cancer therapy. Continued research, interdisciplinary collaborations, and advancements in nanotechnology will be crucial to overcome these obstacles and fully realize the potential of ferrite-based nanoparticles for improved cancer treatment.

8. Conclusion

Although substantial progress has been achieved in the case of *in vivo* applications, to date, confirmed clinical studies are still complicated. Various basic issues such as the size, controlled synthesis, stability, biocompatibility, drug release capacity in physiological conditions, drug-MNMs binding, and others need to be solved. In conclusion, the potential of ferrite-based nanoparticles for improved cancer therapy has shown promising progress in recent years, but several challenges lie ahead for their practical application. The use of ferrite nanoparticles in cancer therapy offers several advantages. These nanoparticles possess unique magnetic properties, such as high magnetic saturation and superparamagnetism, which allow them to be easily manipulated using external magnetic fields. This characteristic enables targeted drug delivery and localized hyperthermia, leading to enhanced therapeutic outcomes while minimizing side effects. Additionally, the surface of ferrite nanoparticles can be modified to attach targeting ligands, antibodies, or drugs, enabling specific targeting of cancer cells and reducing off-target effects. Recent research has demonstrated the efficacy of ferrite-based nanoparticles in various cancer treatment strategies. Magnetic hyperthermia, for instance, utilizes the heating effect generated by ferrite nanoparticles under an alternating magnetic field to selectively kill cancer cells.

Moreover, the ability to load and deliver chemotherapeutic drugs directly to tumors through magnetic targeting has shown improved drug efficacy and reduced systemic toxicity. However, several challenges need to be addressed before ferrite-based nanoparticles can be widely implemented in cancer therapy. One significant obstacle is the limited understanding of the complex interactions between nanoparticles and biological systems. The nanoparticles' behavior *in vivo*, including their biodistribution, clearance, and potential long-term toxicity, requires comprehensive investigation. Further studies are necessary to assess the safety and biocompatibility of these nanoparticles. Ferrite-based nanoparticles hold immense potential for improved cancer therapy. Their unique magnetic properties, coupled with targeted drug delivery and hyperthermia capabilities, offer new avenues for personalized and effective treatment. However, addressing the remaining challenges, such as biocompatibility, synthesis scalability, and regulatory considerations, is crucial for realizing the full clinical potential of these nanoparticles in cancer therapy. Continued research and collaborative efforts are needed to overcome these obstacles and pave the way for the successful implementation of ferrite-based nanoparticles in cancer treatment.

Author details

Jnanranjan Panda[1*], Bhabani Sankar Satapathy[2], Abhishek Mishra[2], Biswabhusan Biswal[2] and Pralaya Kumar Sahoo[2]

1 Faculty of Science, Sri Sri University, Cuttack, India

2 School of Pharmaceutical Sciences, Siksha 'O' Anusandhan (Deemed to be University), Bhubaneswar, India

*Address all correspondence to: jnanranjan.p@srisriuniversity.edu.in

IntechOpen

© 2023 The Author(s). Licensee IntechOpen. This chapter is distributed under the terms of the Creative Commons Attribution License (http://creativecommons.org/licenses/by/3.0), which permits unrestricted use, distribution, and reproduction in any medium, provided the original work is properly cited.

References

[1] Mohapatra J, Xing M, Liu JP. Inductive thermal effect of ferrite magnetic nanoparticles. Materials. 2019;**12**:3208

[2] Rao CN, Muller A, Cheetham AK, editors. The Chemistry of Nanomaterials. Vol. 2. Weinheim: Wiley-VCH; 2004

[3] Hola K, Markova Z, Zoppellaro G, Tucek J, Zboril R. Tailored functionalization of iron oxide nanoparticles for MRI, drug delivery, magnetic separation and immobilization of biosubstances. Biotechnology Advances. 2015;**33**:1162-1176

[4] Durgalakshmi D, Balakumar S, Rajendran S, Naushad M. Functional nanomaterial in energy and environmental science. Nanomaterials for Sustainable Energy and Environmental Remediation. 2020. pp. 1-23

[5] Liu Y, Zhou G, Liu K, Cui Y. Design of complex nanomaterials for energy storage: Past success and future opportunity. Accounts of Chemical Research. 2017;**50**(12):2895-2905

[6] Shabatina TI, Vernaya OI, Shabatin VP, Melnikov MY. Magnetic nanoparticles for biomedical purposes: Modern trends and prospects. Magnetochemistry. 2020;**6**:30

[7] Issa B, Obaidat IM, Albiss BA, Haik Y. Magnetic nanoparticles: Surface effects and properties related to biomedicine applications. International Journal of Molecular Sciences. 2013;**14**(14):21266-21305

[8] Lu AH, Salabas EL, Schuth F. Magnetic nanoparticles: Synthesis, protection, functionalization, and application. Angewandte Chemie. 2007;**46**:1222-1244

[9] Morisako M, Matsumoto MN. Sputtered hexagonal Ba-ferrite films for high-density magnetic recording media. Journal of Applied Physics. 1996;**79**:4881-4883

[10] Bedanta S, Kleeman W. Supermagnetism. Journal of Physics D: Applied Physics. 2009;**42**:013001

[11] Jauhar S, Kaur J, Goyal A, Singhal S. Tuning the properties of cobalt ferrite: A road towards diverse applications. RSC Advances. 2016;**6**:97694-97719

[12] Ghaemi N, Madaeni S, Daraei P, Rajabi H, Zinadini S. Polyethersulfone membrane enhanced with iron oxide nanoparticles for copper removal from water: Application of new functionalized Fe_3O_4 nanoparticles. International Journal of Chemical Engineering. 2015;**263**:101-112

[13] Xu P, Zeng G, Huang D, Hu S, Feng C, Lai C, et al. Synthesis of iron oxide nanoparticles and their application in Phanerochaete chrysosporium immobilization for Pb(II) removal. Colloids and Surfaces A: Physicochemical and Engineering Aspects. 2013;**419**:147-155

[14] Weissleder R, Elizondo G, Wittenberg J, Rabito CA, Bengele HH, Josephson L. Ultrasmall superparamagnetic iron oxide: Characterization of a new class of contrast agents for MR imaging. Radiology. 1990;**175**:489-493

[15] Panda J, Satapathy BS, Mandal B, Sen R, Mukherjee B, Sarkar R, et al. Anticancer potential of docetaxel-loaded cobalt ferrite nanocarrier: An in vitro study on MCF-7 and MDA-MB-231 cell lines. Journal of Microencapsulation. 2020;**38**:36-46

[16] Qiao R, Yang C, Gao M. Superparamagnetic iron oxide nanoparticles: From preparations to in vivo MRI applications. Journal of Materials Chemistry. 2009;**19**(35):6274-6293

[17] Lee N, Hyeon T. Designed synthesis of uniformly sized iron oxide nanoparticles for efficient magnetic resonance imaging contrast agents. Chemical Society Reviews. 2012;**41**:2575

[18] Na HB, Song IC, Hyeon T. Inorganic nanoparticles for MRI contrast agents. Advanced Materials. 2012;**21**:2133-2148

[19] Panda J, Satapathy BS, Majumder S, Sarkar R, Mukherjee B, Tudu B. Engineered polymeric iron oxide nanoparticles as potential drug carrier for targeted delivery of docetaxel to breast cancer cells. Journal of Magnetism and Magnetic Materials. 2019;**485**:165-173

[20] Sung H, Ferlay J, Siegel RL, Laversanne M, Soerjomataram I, Jemal A, et al. Global cancer statistics 2020: GLOBOCAN estimates of incidence and mortality worldwide for 36 cancers in 185 countries. CA: A Cancer Journal for Clinicians. 2021;**71**(3):209-249

[21] Mukherjee B, Satapathy BS, Mondal L, Dey NS, Maji R. Potentials and challenges of active targeting at the tumor cells by engineered polymeric nanoparticles. Current Pharmaceutical Biotechnology. 2013;**14**:1250-1263

[22] El-Say KM, El-Sawy HS. Polymeric nanoparticles: Promising platform for drug delivery. International Journal of Pharmaceutics. 2017;**528**:675-691

[23] Xua Y, Shana Y, Cong H, Shena Y, Yu B. Advanced carbon-based nanoplatforms combining drug delivery and thermal therapy for cancer treatment. Current Pharmaceutical Design. 2018;**24**:1-17

[24] Daraee H, Eatemodi A, Abbasi E, Aval S, Kouhi M. Cells, Application of gold nanoparticles in biomedical and drug delivery. Nanomedicine, and Biotechnology. 2016;**44**:410-422

[25] Gupta K, Gupta M. Synthesis and surface engineering of iron oxide nanoparticles for biomedical applications. Biomaterials. 2005;**26**:3995-4021

[26] Laurent S, Mahmoudi M. The falsifiability of the models for the origin of eukaryotes. International Journal of Molecular Epidemiology and Genetics. 2011;**2**:367-390

[27] Pourjavadi A, Amin SS, Hosseini SH. Delivery of hydrophobic anticancer drugs by hydrophobically modified alginate based magnetic nanocarrier. Industrial and Engineering Chemistry Research. 2018;**57**:822-832

[28] Widder K, Senyei AE, Scarpelli DG. Magnetic microspheres: A model system for site specific drug delivery in vivo. In: Proceedings of the Society for Experimental Biology and Medicine. 1978;**158**:141-146

[29] Senyei A, Widder K, Czerlinski G. Magnetic guidance of drug-carrying microspheres. Journal of Applied Physics. 1978;**49**:3578-3583

[30] Mai T, Fernandes S, Balakrishnan PB, Pellegrino T. Nanosystems based on magnetic nanoparticles and thermo- or pH-responsive polymers: An update and future perspectives. Accounts of Chemical Research. 2018;**51**:999-1013

[31] Williams H. The application of magnetic nanoparticles in the

treatment and monitoring of cancer and infectious diseases. Bioscience Horizons. 2017;**10**:1-10

[32] Lazaro FJ, Abadia AR, Romero MS, Gutierrez L, Lazaro J, Morales MP. Magnetic characterisation of rat muscle tissues after subcutaneous iron dextran injection. Biochimica et Biophysica Acta (BBA) - Molecular Basis of Disease. 2004;**1740**:434-445

[33] Lubbe AS, Alexiou C, Bergemann C. Clinical applications of magnetic drug targeting. The Journal of Surgical Research. 2001;**95**:200-206

[34] Umut E. Surface modification of nanoparticles used in biomedical applications. Modern Surface Enginnering Treatments. 2013;**20**:185-208

[35] Gupta K, Wells S. Surface-modified superparamagnetic nanoparticles for drug delivery: Preparation, characterization, and cytotoxicity studies. IEEE Transactions on Nanobioscience. 2004;**3**:66-73

[36] Patil RM, Shete PB, Thorat ND, Otari SV, Barick KC, Prasad A, et al. Non-aqueous to aqueous phase transfer of oleic acid coated iron oxide nanoparticles for hyperthermia application. RSC Advances. 2014;**4**:4515

[37] Yadavalli T, Jain H, Chandrasekharan G, Chennakesavulu R. Magnetic hyperthermia heating of cobalt ferrite nanoparticles prepared by low temperature ferrous sulfate based method. AIP Advances. 2016;**6**:055904

[38] Ansari SM, Bhor RD, Pai KR, Mazumder S, Sen D, Kolekar YD, et al. Size and chemistry controlled cobalt-ferrite nanoparticles and their anti-proliferative effect against the MCF-7 breast cancer cells. ACS Biomaterials Science and Engineering. 2016;**2**:2139-2152

[39] Shah SA, Majeed A, Rashid K, Awan S. PEG-coated folic acid-modified superparamagnetic $MnFe_2O_4$ nanoparticles for hyperthermia therapy and drug delivery. Materials Chemistry and Physics. 2013;**138**:703e708

[40] Wang G, Zhao D, Ma Y, Zhang Z, Che H, Mu J, et al. Synthesis and characterization of polymer-coated manganese ferrite nanoparticles as controlled drug delivery. Applied Surface Science. 2018;**428**:258-263

[41] Maiti D, Saha A, Devi PS. Surface modified multifunctional $ZnFe_2O_4$ nanoparticles for hydrophobic and hydrophilic anti-cancer drug molecule loading. Physical Chemistry Chemical Physics. 2016;**18**:1439-1450

[42] Pan UN, Sanpui P, Paul A, Chattopadhyay A. Surface-complexed zinc ferrite magnetofluorescent nanoparticles for killing cancer cells and single-particle-level cellular imaging. ACS Applied Nano Materials. 2018;**6**:2496-2502

[43] Majumder S, Dey S, Bagani K, Dey SK, Banerjee S, Kumar S. A comparative study on the structural, optical and magnetic properties of Fe_3O_4 and Fe_3O_4@SiO_2 core–shell microspheres along with an assessment of their potentiality as electrochemical double layer capacitors. Dalton Transactions. 2015;**44**:7190-7202

[44] Mondal R, Sarkar K, Dey S, Majumdar D, Bhattacharya SK, Sen P, et al. Magnetic, pseudocapacitive, and H_2O_2-electrosensing properties of self-assembled superparamagnetic $CoO.3ZnO.7Fe_2O_4$ with enhanced saturation magnetization. ACS Omega. 2019;**4**:12632-12646

[45] Hadjipanayis GC, Siegel RW. Nanophase Materials: Synthesis - Properties - Applications. Dordrecht, Boston: Kluwer Academic Publishers; 1993

[46] Sjoegren CE, Briley-Saeboe K, Hanson M, Johansson C. Magnetic characterization of iron oxides for magnetic resonance imaging. Official Journal of the Society of Magnetic Resonance in Medicine/Society of Magnetic Resonance in Medicine. 1994;**31**:268-272

[47] Cornell RM, Schwertmann U. The iron oxides: Structure, properties, reactions, occurrences, and uses. Mineralogical Magazine. 1997;**61**(408):740-741

[48] Cotton FA. Advanced Inorganic Chemistry. New York: Wiley; 1988

[49] Euliss LE, Grancharov SG, Brien SO, Deming TJ, Stucky GD, Murray CB, et al. Cooperative assembly of magnetic nanoparticles and block copolypeptides in aqueous media. Nano Letters. 2003;**3**:1489

[50] Liu X, Guan Y, Ma Z, Liu H. Surface modification and characterization of magnetic polymer nanospheres prepared by miniemulsion polymerization. Langmuir. 2004;**20**:10278

[51] Suk JS, Xua Q, Kima N, Hanesa J, Ensigna LM. PEGylation as a strategy for improving nanoparticle-based drug and gene delivery. Advanced Drug Delivery Reviews. 2016;**99**:28-51

[52] Lubbe S, Bergemann C, Huhnt W, Fricke T, Riess H, Brock JW, et al. Preclinical experiences with magnetic drug targeting: Tolerance and efficacy. Cancer Research. 1996;**56**:4694-4701

[53] Kohler N, Sun C, Wang J, Zhang M. Methotrexate-modified superparamagnetic nanoparticles and their intracellular uptake into human cancer cells. Langmuir. 2005;**21**(19):8858-8864. DOI: 10.1021/la0503451

[54] Mohapatra S, Rout SR, Maiti S, Maiti TK, Panda AB. Monodisperse mesoporous cobalt ferrite nanoparticles: Synthesis and application in targeted delivery of antitumor drugs. Journal of Materials Chemistry. 2011;**21**(25):9185-9193

[55] Georgiadou V, Kokotidou C, Le Droumaguet B, Carbonnier B, Choli-Papadopoulou T, Dendrinou-Samara C. Oleylamine as a beneficial agent for the synthesis of $CoFe_2O_4$ nanoparticles with potential biomedical uses. Dalton Transactions. 2014;**43**:6377-6388

[56] Cotica LF, Freitas VF, Silva DM, Honjoya K, Honjoya K, Santos IA, et al. Thermal decomposition synthesis and assessment of effects on blood cells and in vivo damages of cobalt ferrite nanoparticles. Journal of Nano Research. 2014;**28**:131-140

[57] Matsuda S, Nakanishi T, Kaneko K, Osaka T. Synthesis of cobalt ferrite nanoparticles using spermine and their effect on death in human breast cancer cells under an alternating magnetic field. Electrochimica Acta. 2015;**183**:153-159

[58] Wang G, Zhou F, Li X, Li J, Ma Y, Mu J, et al. Controlled synthesis of L-cysteine coated cobalt ferrite nanoparticles for drug delivery. Journal of Magnetism and Magnetic Materials. 2018;**44**:13588-13594

[59] Fan H, Li B, Shi Z, Zhao L, Wang K, Qiu D. A fibrous morphology silica-$CoFe_2O_4$ nanocarrier for anti-cancer drug delivery. Ceramics International. 2018;**44**:2345-2350

[60] Gaihrea B, Khil MS, Lee DR, Kim HY. Gelatin-coated magnetic iron oxide nanoparticles as carrier system: Drug loading and in vitro drug release study. International Journal of Pharmaceutics. 2009;**365**:180-189

[61] Ling Y, Wei K, Luo Y, Gao X, Zhong S. Dual docetaxel/superparamagnetic iron oxide loaded nanoparticles for both targeting magnetic resonance imaging and cancer therapy. Biomaterials. 2011;**32**:7139-7150

[62] Schleich N, Sibret P, Danhier P, Ucakar B, Laurent S, Muller RN, et al. Dual anticancer drug/superparamagnetic iron oxide-loaded PLGA-based nanoparticles for cancer therapy and magnetic resonance imaging. International Journal of Pharmaceutics. 2013;**447**:94-101

[63] Rudzka K, Viota JL, Munoz-Gamez JA, Carazo A, Ruiz-Extremera A, Delgadoa ÁV. Nanoengineering of doxorubicin delivery systems with functionalized maghemite nanoparticles. Colloids and Surfaces, B: Biointerfaces. 2013;**111**:88-96

[64] Zhou D, Sun Y, Zheng Y, Ran H, Li P, Wang Z, et al. Superparamagnetic PLGA–iron oxide microspheres as contrast agents for dual-imaging and the enhancement of the effects of high-intensity focused ultrasound ablation on liver tissue. RSC Advances. 2015;**5**:35693

[65] Pham XN, Nguyen TP, Pham TN, Tran TTN, Tran TVT. Synthesis and characterization of chitosan-coated magnetite nanoparticles and their application in curcumin drug delivery. Advances in Natural Sciences: Nanoscience and Nanotechnology. 2016;**7**:045010

[66] Cui Y, Zhang M, Zeng F, Jin H, Xu Q, Huang Y. Dual-targeting magnetic PLGA nanoparticles for codelivery of paclitaxel and curcumin for brain tumor therapy. ACS Applied Materials & Interfaces. 2016;**8**:32159-32169

[67] Mohanta SC, Saha A, Sujatha Devi P. PEGylated Iron oxide nanoparticles for pH responsive drug delivery application. Materials Today: Proceedings. 2018;**5**:9715-9725

[68] Ayyanaara S, Kesavanb MP, Sivaramanc G, Maddiboyinad B, Annaraje J, Rajeshf J, et al. A novel curcumin-loaded PLGA micromagnetic composite system for controlled and pH-responsive drug delivery. Colloids and Surfaces A. 2019;**573**:188-195

[69] Zhu X, Li J, Peng P, Nassab NH, Smith BR. Quantitative drug release monitoring in tumors of living subjects by magnetic particle imaging nanocomposite. Nano Letters. 2019;**19**:6725-6733

[70] Khana S, Setuaa S, Kumaria S, Dana N, Masseya A, Hafeeza BB, et al. Superparamagnetic iron oxide nanoparticles of curcumin enhance gemcitabine therapeutic response in pancreatic cancer. Biomaterials. 2019;**208**:83-97

[71] Amiri M, Salavati-Niasari M, Akbari A. Magnetic nanocarriers: Evolution of spinel ferrites for medical applications. Advances in Colloid and Interface Science. 2019;**265**:29-44

[72] Liu X, Zhang Y, Wang Y, Zhu W, Li G, Ma X, et al. Comprehensive understanding of magnetic hyperthermia for improving antitumor therapeutic efficacy. Theranostics. 2020;**10**(8):3793

[73] Lima-Tenorio MK, Pineda EA, Ahmad NM, Fessi H, Elaissari A. Magnetic nanoparticles: In vivo cancer diagnosis and therapy. International Journal of Pharmaceutics. 2015;**493**(1-2):313-327

Chapter 5

Correlation between Low Field Microwave Power Absorption and Soft Magnetic Properties of Ferrites

Herlinda Montiel, José Francisco Barrón-López and Guillermo Alvarez

Abstract

We describe studies on ferrites that use the Low Field Absorption (LFA) technique. These are made in soft magnetic polycrystalline ferrites magnetite, garnet iron, and nickel-zinc ferrite, where we show LFA spectra, highlighting their main characteristics and illustrating how magnetic phase transitions and magnetic anisotropy are detected employing these measurements. In this chapter, we also present a review of ferromagnetic resonance (FMR) studies in ferrites. Ferromagnetic materials exhibit various behaviors when subjected to ac magnetic fields, such as domain wall relaxation (DWR) and FMR. For instance, most ferrites generally find DWR at relatively low frequencies, below 10 MHz, while FMR must satisfy the Larmor equation. Microwave power absorption can be associated with magnetization processes as a function of dc fields, such as LFA and FMR. Microwave absorption centered at zero magnetic fields has been observed in various materials: superconductors, ferrites, and semiconductors. For ferrites, the LFA signal with the same phase of the ferromagnetic resonance follows the thermal behavior of the magnetization. For magnetically ordered materials, the LFA signal is evidence of the existence of a ferromagnetic phase, this signal is not present in the paramagnetic state, and it appears as the temperature is decreased below Curie temperature.

Keywords: soft ferrites, low field absorption (LFA), ferromagnetic resonance (FMR), phase transition, magnetically modulated microwave absorption spectroscopy (MAMMAS)

1. Introduction

Ferrites are a group of magnetic ceramic oxides containing the ferric ion Fe^{3+} in their structure and a divalent (M^{2+}) or trivalent rare earth (R^{3+}) cation. Accordingly, to their crystal structure, ferrites can be classified as spinels (MFe_2O_4), garnets ($R_3Fe_5O_{12}$), perovskites ($RFeO_3$), or as hexagonal structures related to the magnetoplumbite mineral ($MO{\cdot}6Fe_2O_3$) [1, 2]. This diversity in composition and crystal structure provides a broad spectrum of magnetic and electrical properties which can be tuned and engineered for electronic and biomedical applications.

IntechOpen

This chapter will focus on soft magnetic ferrites such as the spinel [3] and the garnet crystal structure [4, 5]. These ferrites are characterized by their high magnetic permeability, high electrical resistivity, and moderate permittivity from dc to GHz frequencies [6]. Soft ferrites are applied in electronic devices of high frequency [7] due to their ability to interact with dc and ac magnetic fields for transporting information or energy [6, 7] and by shielding electronic components from undesirable electromagnetic interference (EMI) [8].

The applicability of the soft ferrites is associated with the magnetic susceptibility (χ) or its magnetic permeability ($\mu_r = \mu/\mu_0 = 1 + \chi$). As well as soft ferrite is anisotropic, losses, and dispersive media; their magnetic permeability is a complex quantity and depends on the temperature (T), magnetic field frequency (ω), strength and direction (H), and stress, where the characterization of $\mu(\omega)$ as a function of the frequency is fundamental for the development of applications based on soft magnetic materials. Also, the measurement techniques of $\mu(\omega)$ concerning magnetic fields, frequency, and temperature are valuable tools that provide a vision into the magnetic ordering, magnetic couplings, anisotropies, energy losses, and phase transitions in magnetic materials.

This chapter is dedicated to the microwave characterization of the soft magnetic ferrites magnetite (Fe_3O_4), nickel-zinc ferrite (NZFO: $(Ni_x, Zn_{1-x})Fe_2O_4$), and yttrium iron garnet (YIG: $Y_3Fe_5O_{12}$). The focus of this chapter will be on the characterization of these systems and their magnetic anisotropies and phase transitions using the microwave power absorption techniques that historically are known as Ferromagnetic resonance (FMR) [9, 10], Low field absorption (LFA) [11] and magnetically modulated microwave absorption spectroscopy (MAMMAS) [12].

1.1 Soft magnetic materials under alternating magnetic fields

Ferromagnetic materials exhibit various behaviors when subjected to ac magnetic fields, ranging from domain wall relaxation (DWR) to ferromagnetic resonance (FMR). At radio frequency, in the 10^3–10^6 Hz range, the changes in magnetic permeability are mainly associated with the movement of domain walls (DW), where this process can be modeled to assume that DW moves like an elastic membrane driven by an externally applied ac field. In this model, it is assumed that the DW can be elastically deformed by the amount $x(t)$ in response to an ac field $H_0e^{-i\omega t}$; this process is described through the differential equation [1, 13]:

$$m\ddot{x} + \beta\dot{x} + \alpha x = 2M_sH_0e^{-i\omega t} \tag{1}$$

Where m is a *domain wall effective mass* per wall surface area, β is a magnetic *viscous damping factor*, and α is a *restoring constant, similar to a deadened harmonic oscillator*. Eq. (1) can exhibit relaxation or resonance of domain walls depending on the ratios of constants m, α, and β. By defining the resonance $\omega_0 = \sqrt{\alpha/m}$, and relaxation $\omega_r = \alpha/\beta$ frequencies, it is possible to explore the DW dynamics as the damping and restoring parameters vary.

Figure 1 shows the calculated complex inductance ($L^*(\omega)$) spectra for ferrite samples characterized by different ratios of relaxation to resonance frequencies. Inductance is directly proportional to magnetic permeability $L^*(\omega) = G_0\mu^*(\omega)$, where G_0 is a shape-dependent geometrical factor. This behavior goes from a relaxation-driven process toward a resonance-dominated one as the relaxation frequency becomes larger than the resonance one.

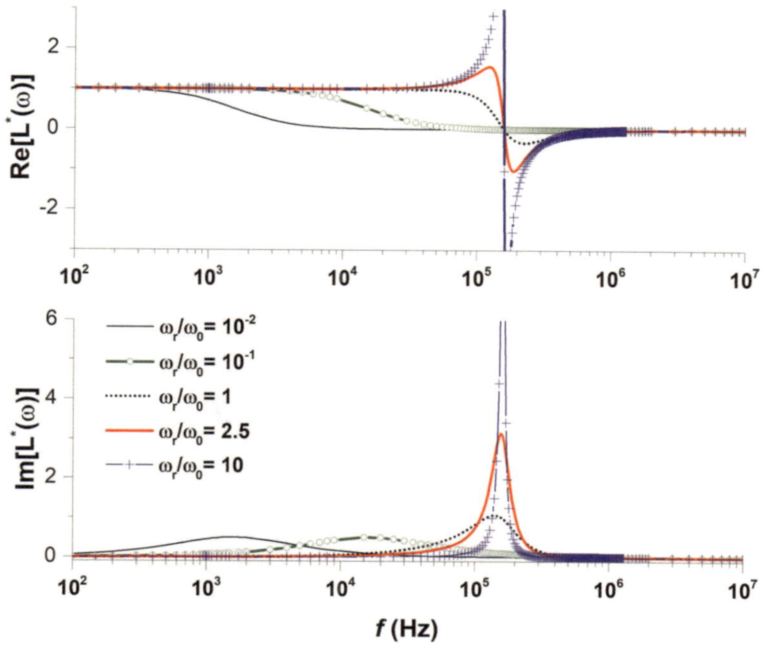

Figure 1.
Calculated complex inductance $L^*(\omega)$ spectra for ferrites with different ratios of relaxation to resonance frequencies.

1.2 Microwave power absorption techniques

The volumetric density of the microwave power absorbed by a magnetic sample is directly related to the electrodynamic properties of the material and is given as [14, 15]:

$$\frac{dP}{dV} = \frac{1}{2}\left(\sigma'|E_{ac}|^2 + \omega\varepsilon''|E_{ac}|^2 + \omega\mu''|H_{ac}|^2\right) \tag{2}$$

Where ω is the angular frequency of the electromagnetic radiation, ε'' is the imaginary part of the electric permittivity, and μ'' is the imaginary part of the magnetic permeability. Each term at the right side of the equation corresponds to losses due to the Joule effect, dielectric, and magnetic mechanisms, respectively.

Microwave power absorption can be measured as a function of the radiation frequency (ω) with the DC magnetic field (H_{DC}) constant; or as a function of H_{DC} under a fixed microwave frequency. The first kind of measurement, broadband FMR [16, 17], requires a microstrip line or coplanar waveguide setting for transporting the microwave field. In this setup, the sample is mounted onto the waveguide, and the reflected and absorbed waves are analyzed while the system is kept under a fixed DC magnetic field H_{DC}.

In the second type of FMR setup [17, 18], the sample is mounted inside and in the center of a resonant cavity coupled to a microwave source through a magic-T bridge. The Microwave absorption is obtained by analyzing the reflected wave from the sample-loaded cavity due to a change in the quality factor (Q), and later it is guided to

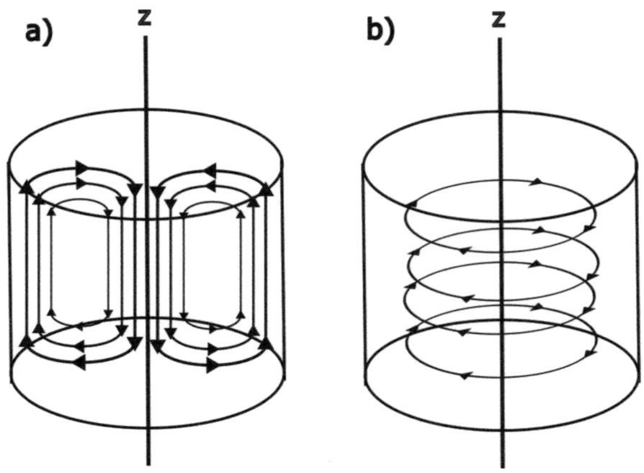

Figure 2.
Schematic distribution of (a) magnetic and (b) electric field lines inside a cylindrical TE_{011} microwave cavity. Adapted from [20].

a crystal detector using a microwave circulator. These microwave measurements are performed under a fixed frequency while varying the DC magnetic field (H_{DC}). It is necessary to mention that as the resonant cavity and the waveguide are designed for specific microwave bands, e.g., X-band (8.8–9.8 GHz), they only allow for a small frequency variation, meaning that they need to be changed each time a different measuring frequency is required.

The kind of microwave absorption techniques employed in this chapter (FMR, LFA, and MAMMAS) were implemented with resonant cavities in the TE_{011} mode [12, 17, 19], where the sample is placed at a maximum magnetic field (H_{ac}) and a minimum electric field (E_{ac}), see **Figure 2**. This setup holds that $|E_{ac}| << |''H_{ac}|$, so the magnetic losses dominate over the dielectric losses. Also, for most non-metallic magnetic samples, the electrical conductivity is minimal; therefore, the Joule effect losses are minimal. So, the main contribution comes from magnetic losses, reflecting the high-frequency collective response of the ordered electron spins on the material.

Most of the results discussed in this chapter were obtained in an electron paramagnetic resonance (EPR) spectrometer JEOL JES-RES3X (see **Figure 3**). It is also necessary to mention that microwave absorption is modulated by a magnetic field (H_{mod}) of 100 kHz and is generated by Helmholtz coils on each side of the cavity. Therefore, the spectrometer signal is detected through the method of detection-homodyne with lock-in amplification, i.e., the output signal of the spectrometer (S_{FMR}) is proportional to the magnetic field derived from the microwave power absorption of the sample at a given microwave frequency (ω), modulation field (H_{mod}) and temperature (T) conditions [21]:

$$S_{FMR} = \left(\frac{dP}{dH}\right)_{\omega, H_{mod}, T} \tag{3}$$

1.2.1 Ferromagnetic resonance (FMR)

The ferromagnetic resonance (FMR) phenomenon is the resonant absorption of microwave energy due to the Larmor precession of the magnetic moment in a material (**Figure 4**). The Larmor frequency gives FMR condition [9, 10, 22]:

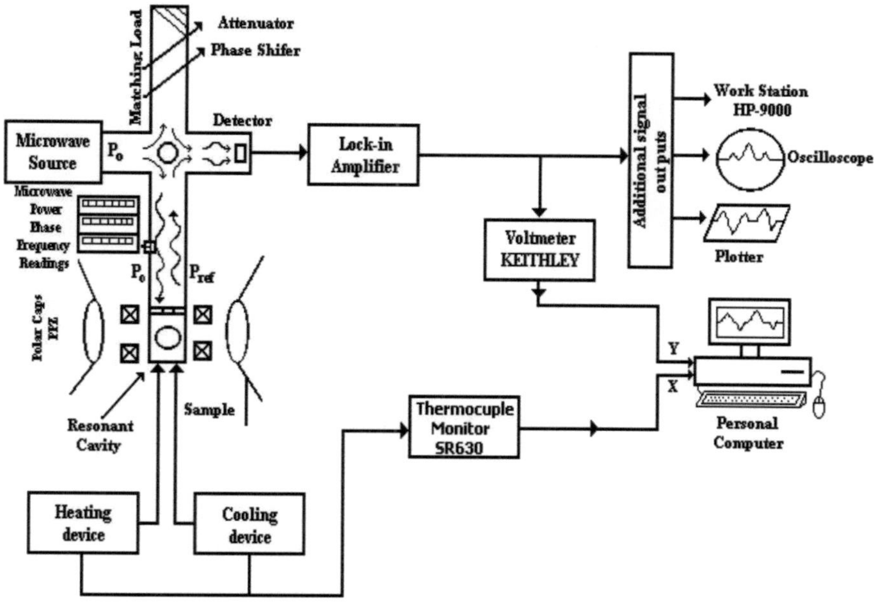

Figure 3.
Block diagram for implementing FMR, LFA, and MAMMAS techniques in a JEOL JES-RES3X spectrometer at X-band. Adapted from [12, 20].

$$\omega_0 = \gamma\mu_0 H_0 \tag{4}$$

Here, $\gamma = \frac{g\mu_B}{\hbar} = 1.76086 \times 10^{11}\,\frac{\text{rad}}{\text{s}\bullet\text{T}}$ ($\gamma/2\pi = 28.25$ GHz/T) is the *gyromagnetic ratio* of the electrons, g is the Landé-factor of the electrons, and H_0 is the *resonance field* of the system.

The magnetization process at microwave frequencies can be modeled via the Landau- Lifshitz-Gilbert (LLG) equation [23]:

$$\frac{d\vec{M}}{dt} = -\gamma\mu_0\vec{M} \times \vec{H}_0 + \frac{\alpha_G}{M}\left(\vec{M} \times \frac{d\vec{M}}{dt}\right) \tag{5}$$

Here, α_G is the dimensionless Gilbert damping parameter, which describes the relaxation of magnetization during Larmor precession.

LLG equation describes the dynamical response and relaxation of magnetization under given conditions of dc (H_{DC}) and ac (H_{ac}) fields. Because of the magnetic ordering, the effective field experienced by the magnetization is given by the superposition of the applied DC field and the internal field of the material:

$$H_0 = H_{DC} + H_{int} \tag{6}$$

The internal field (H_{int}) represents the restoring torque effects experienced by the magnetic moment $\hat{\boldsymbol{m}}$ due to crystal anisotropy(H_K), demagnetizing fields (H_D), exchange coupling interaction (H_J), and magnetoelastic effects (H_λ). Therefore, the changes in the resonance field are due to sample orientation, chemical composition,

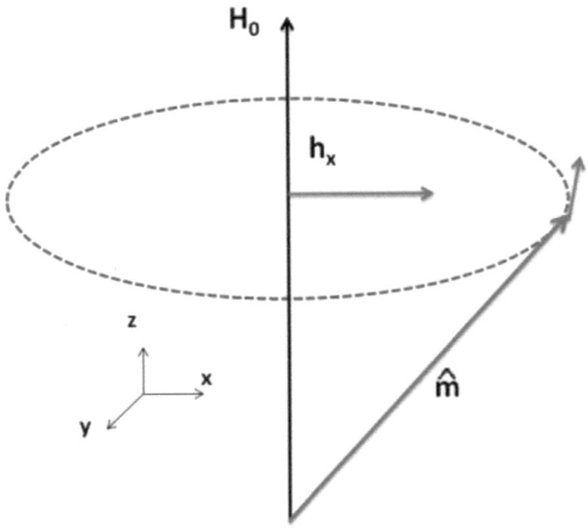

Figure 4.
Schematic representation of the Larmor precession. A magnetic moment $\hat{\mathbf{m}}$ describes a precession cone around the magnetic field $\mathbf{H_o}$ and absorbs the transverse microwave field $\mathbf{h_x}$. Adapted from [9].

microstructure, and temperature, providing rich information about the internal field of magnetic materials.

Also, FMR experiments allow for obtaining dynamical information regarding the magnetic relaxation of materials via the analysis of the FMR linewidth.

It can be shown that, near the resonance field, the real (χ') and imaginary (χ'') components of the magnetic susceptibility can be approximately described by the Lorentzian resonance curves [22]:

$$\chi' \approx \left(\frac{\gamma M_s}{2\alpha_G \omega}\right)\frac{\beta}{1+\beta^2} \tag{7}$$

$$\chi'' \approx \left(\frac{\gamma M_s}{2\alpha_G \omega}\right)\frac{1}{1+\beta^2} \tag{8}$$

With the parameter $\beta = \frac{H-(\omega/\gamma)}{\alpha\omega}$ for FMR measurement at a constant frequency. ΔH_{FWHM} of the imaginary part (χ'') is related to the Gilbert damping parameter by the relation [22]:

$$\Delta H_{FWHM} = \frac{2\alpha_G \omega}{\gamma} \tag{9}$$

Eqs. (7)–(9) indicate that a smaller Gilbert damping factor makes the microwave absorption peak value more intense and concentrated in a small field interval. Else, as α_G increases, the absorption takes place into a wider field interval with a lower peak value.

1.2.2 Low field absorption (LFA)

LFA technique refers to the microwave absorption observed around zero DC magnetic field within a typical field width of $\mu_0 H_{DC} = \pm 100\text{mT}$. LFA measurements are performed in the typical microwave cavity setup for electron paramagnetic resonance (EPR) spectroscopy (**Figure 3**). EPR spectrometer is equipped with a zero-field cross-unit which compensates for the remanence of the electromagnet, then allows precise measurements at small field values near zero [11, 20].

The typical LFA signal may exhibit hysteresis, observed in superconducting and magnetically ordered materials [11, 12]. So, the LFA signal is associated with the absorption dynamics of microwaves for fluxoid tubes in High-T_c superconductors and the interaction of microwaves with magnetic moments in the magnetic domains for ferromagnetic materials [12, 24].

LFA signal only appears below the corresponding Curie temperature T_C, a signature of the collective spin ordering in magnetic materials [25–28]. In magnetic materials, the hysteresis in the LFA signal has been associated with magnetization processes occurring in the magnetic domains of the non-saturated state [29]. Experiments on nickel samples [29] and amorphous cobalt-rich ribbons [30] also suggest that the LFA signal is heavily correlated with materials anisotropy and demagnetizing fields.

1.2.3 Magnetically modulated microwave absorption spectroscopy (MAMMAS)

Magnetically modulated microwave absorption spectroscopy (MAMMAS) [11, 12], or MFMMS (Magnetic field modulated microwave spectroscopy) as it is called in other works [15], is the study of the losses of the microwave power absorption as temperature function at a fixed magnetic field value (H_{DC}). So, the MAMMAS signal is the magnetic field derived from the microwave power absorption, measured at a fixed magnetic field (H_{DC}), as a temperature function:

$$\text{MAMMAS} = \left(\frac{dP}{dH}\right)_{\omega, H_{DC}} (T) \tag{10}$$

It is necessary to mention that the FMR and LFA spectra are measured at a fixed temperature. However, the MAMMAS signal can be measured under a continuous temperature scan. Therefore, the MAMMAS technique is ideal for studying phase transitions in magnetic materials [15, 20]. The high sensitivity of this technique to detect the magnetic phase transitions is associated with the changes in electromagnetic properties of the materials with long-range electronic ordering. As this ordering is destroyed by thermal energy, marked changes exist in the electrical conductivity, permittivity, and permeability of materials during a phase transition. Finally, according to Eq. (2), changes in these properties affect the microwave power absorption of the magnetic medium.

2. Microwave power absorption studies in soft magnetic ferrites

Microwave power absorption techniques like FMR, LFA, and MAMMAS have been applied for various studies about magnetic ordering, phase transitions, and anisotropy characterization on soft magnetic ferrites. Here, we present a survey on the studies carried out in magnetite, nickel-zinc ferrite (NZFO), and yttrium iron garnet (YIG) employing these techniques.

2.1 Magnetite (Fe_3O_4)

Magnetite (Fe_3O_4) is the natural form of iron oxide and is considered a typical ferrite material. It crystallizes in the inverse spinel structure, whose general formula is represented by AB_2O_4. This structure belongs to the high symmetry space group $Fd\bar{3}m$ (No. 227), and the unit cell has a lattice parameter of $a_0 = 8.3967A$ [31]. The spinel structure is formed by a closed cubic packing of oxygen anions with cations distributed between tetrahedral or A-sites and octahedral or B-sites. In a normal spinel, the divalent cation occupies the A-sites, while trivalent cations occupy octahedral B-sites as represented by: (A^{2+}) $[B^{3+}{}_2]O_4$. However, the magnetite is an inverse spinel, where the A-sites are occupied by half of the ferric ions while the B-sites contain the other half of Fe^{3+} ions and the ferrous Fe^{2+} cations. Then, cationic distribution in magnetite corresponds to (Fe^{3+}) $[Fe^{3+}Fe^{2+}]O_4$.

Magnetite has a low-temperature phase transition known as the Verwey transition [32–35], where the magnetite has a low crystal symmetry, i.e., goes changing from the cubic $Fd\bar{3}m$ space group to a monoclinic structure (space-group *Cc*) when cooling. This phase transition is also accompanied by an ordering change phenomenon of the ferric and ferrous ions in octahedral sites, and it is manifested as a sharp increase in the electrical resistivity of the magnetite at $T = T_V \cong 122$ K [33, 36]. In this transition, by diminishing the temperature, the metallic conductive cubic phase changes toward a monoclinic insulator phase. This low symmetry phase (**Figure 5**) is described by a superstructure where the 8 tetrahedra and 16 octahedral sites become non-equivalent among them [36].

The Verwey transition has been studied by different techniques [37], such as electrical resistivity, magnetoresistance [38], infrared and Raman spectroscopes [39], X-Ray diffraction (XRD), magnetometry, susceptibility, Mössbauer spectroscopy, calorimetric, nuclear magnetic resonance (NMR), and X-ray resonant absorption [32, 33, 37]. Also, FMR, LFA, and MAMMAS techniques have been applied to study this phase transition [24, 40].

In an earlier FMR study [41] on magnetite single crystals at X-band and K-band (18–26.5 GHz), the disappearance of the FMR signal has been observed at $T = T_V$ for X-band measurements. Furthermore, it could be interpreted as an increased Larmor frequency far above the X-band frequency range. This increase in FMR frequency is attributed to the larger anisotropy change of the monoclinic phase, which is added to the internal field in the magnetite (Eq. (6)).

More recently, the broadband FMR technique was used for studying the Verwey transition in an epitaxial magnetite film grown on $MgGa_2O_4$ (001) substrates [42]. FMR linewidth and the Gilbert damping showed an abrupt increase below the $T_V = 110$ K, which is consistent with the expected crystal anisotropy change in the [010] direction.

FMR experiments were carried out in magnetite nanopowders, see **Figure 6**, following the Verwey transition through the changes in the resonance field (H_r) and peak-to-peak linewidth (ΔH_{pp}) in these spectra [40], where $\Delta H_{pp}$and H_r have a minimum and an inflection point near *the* T_V (**Figure** 7), respectively.

As is known, in the limit of independent, non-interacting particles, the FMR linewidth is proportional to the anisotropy field, i.e., $\Delta H_{pp} \propto H_K$ [43]. So, the minimum in the FMR linewidth occurring before the Verwey transition can be associated with the isotropic point ($T_{iso} \approx 130$ K [33]) in the magnetocrystalline anisotropy field for the cubic phase of magnetite, where K_1 passes through zero and changes to a positive sign for $T_V < T < T_{iso}$ (**Figure 8**). In this way, *the* ΔH_{pp} minimum occurs due to the

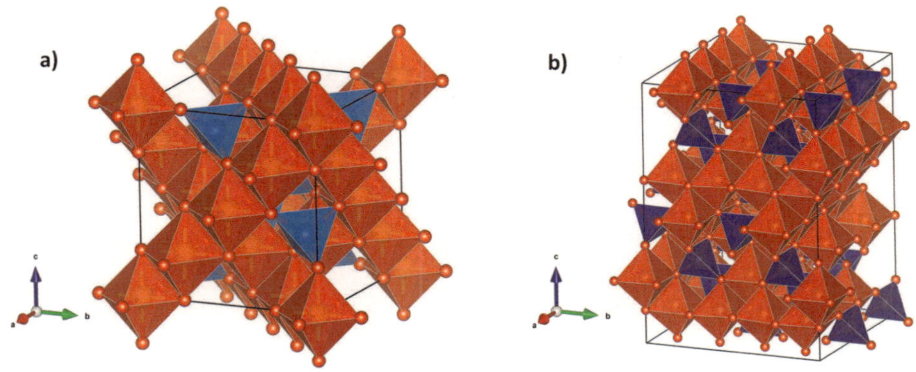

Figure 5.
Crystal structures of magnetite in the a) high symmetry phase (Fd$\bar{3}$m) *for* T > T_V, *and b) in the low symmetry monoclinic phase* (Cc) *for* T < T_V.

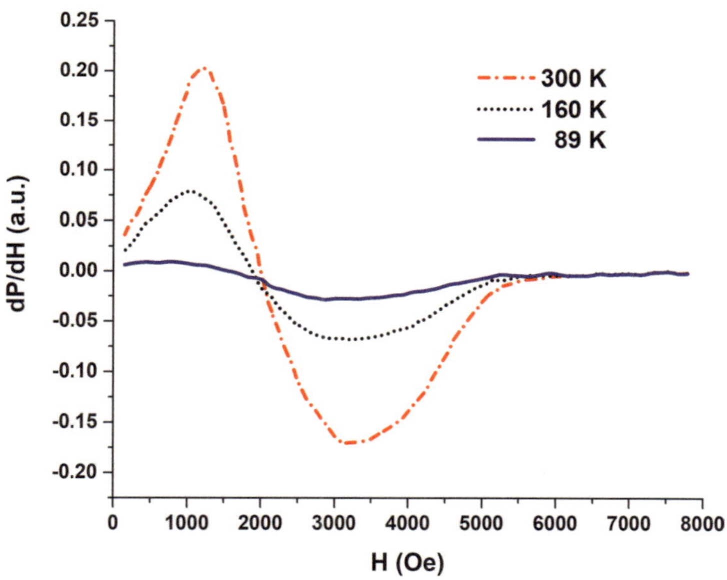

Figure 6.
FMR spectra in magnetite nanopowders for selected temperatures. Adapted from [40].

nearly zero crystalline anisotropy in magnetite before the Verwey transition, and the residual linewidth reflects the contribution of crystalline defects, electric losses, and dipolar coupling in the nanopowders.

LFA and MAMMAS techniques have been applied for studying the Verwey transition in bulk and nanopowders magnetite samples. **Figure 9a** shows LFA spectra in bulk magnetite samples at several temperatures. The linewidth (ΔH_{LFA}) and the enclosed area (A_{LFA}) of the LFA signal showed a very similar temperature dependence. ΔH_{LFA} showed a minimum of 126 K, which was associated with the minimum in anisotropy. On further cooling, ΔH_{LFA} begins to grow again due to the increase in anisotropy of the monoclinic phase.

In nanoparticles [40], LFA spectra can be described by a superposition of two components, one with hysteresis convoluted with a linear one. As is shown in

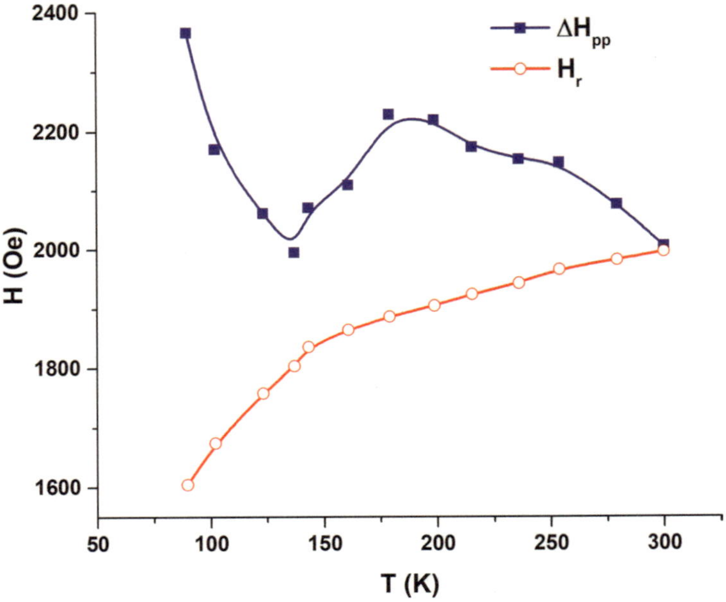

Figure 7.
Temperature dependence of the resonance field (H_r) and peak-to-peak linewidth (ΔH_{pp}) in magnetite nanopowders. Adapted from [40].

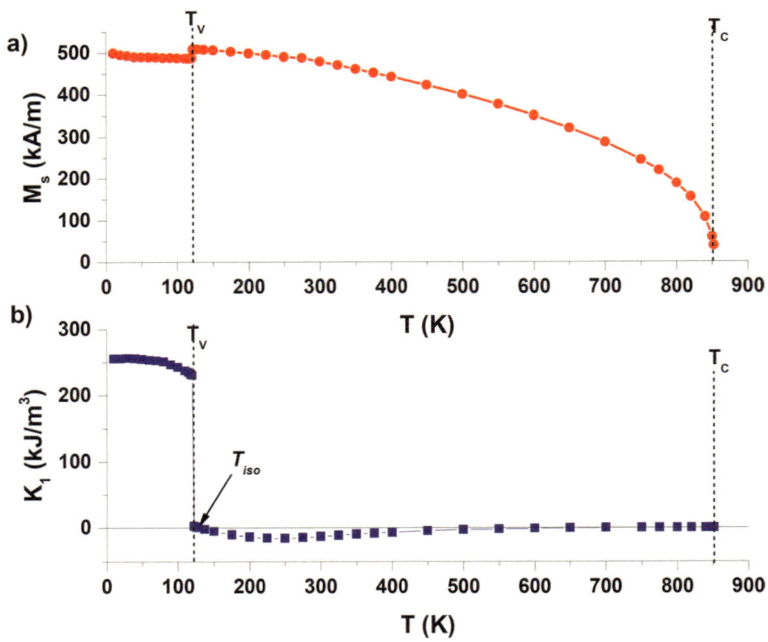

Figure 8.
Temperature dependence of (a) the saturation magnetization (filled circles) and (b) the lead magneto term of the magnetocrystalline anisotropy constant (K_1-filled squares). Data were interpolated from reference [33].

Figure 9b, the slope of the linear component increases with temperature, showing a significant step at T > 137 K, which has been associated with the onset of the change in electron dynamics in B-sites.

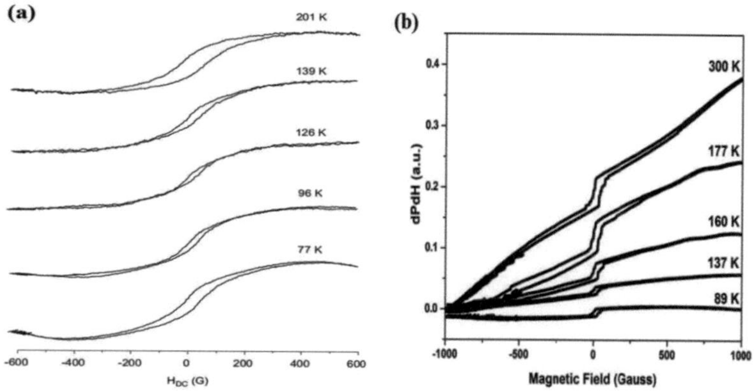

Figure 9.
LFA spectra in (a) bulk and (b) nanopowders magnetite at different temperatures. Adapted from [24, 40], respectively.

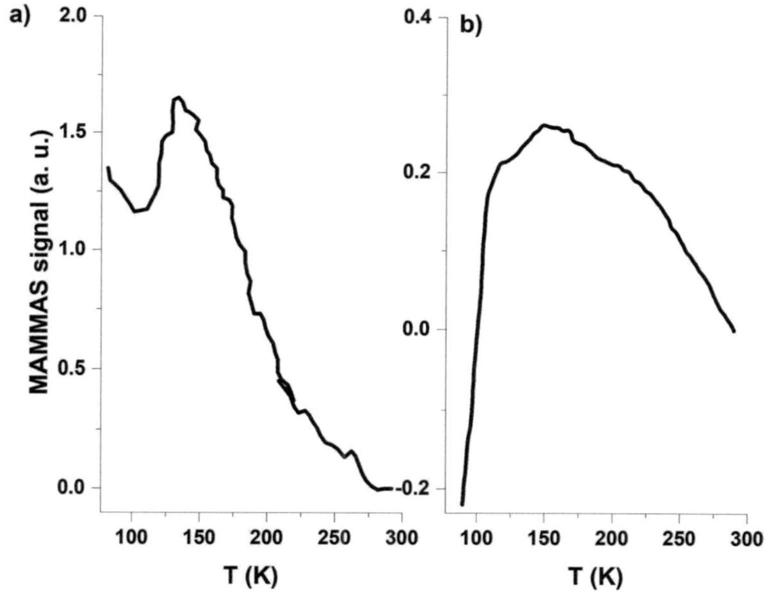

Figure 10.
MAMMAS response for (a) bulk and (b) nanopowders magnetite samples. Adapted from [24, 40], respectively.

MAMMAS spectra for bulk [24] and nanopowders [40] for magnetite samples are shown in **Figure 10a** and **b**, respectively. In both MAMMAS spectra, an increment in the microwave absorption is observed, due to an increment in the absorbing centers' quantity and which produces an increase in the magnetoconductance of the samples, until reaching a maximum value. Later, the MAMMAS spectra diminish, and this behavior is associated with a decrease in the quantity of absorbing centers by the process of antiparallel spin alignment. It is necessary to mention that in the bulk sample, MAMMAS response increases to low temperatures due to an increase of the magnetocrystalline anisotropy in the magnetite monoclinic phase.

Wampler et al. [15] used the MAMMAS technique for studying the Verwey transition in magnetite epitaxial film growth onto MgO (001) substrates. They also provided electromagnetic modeling to explain the changes in microwave absorption for magnetite epitaxial films as a temperature function. In their model, the changes in MAMMAS spectra (renamed as magnetic field modulated microwave spectroscopy, MFMMS) are associated with changes in the magnetic moment and the magnetoresistive behavior of magnetite. MAMMAS measurements were performed under different DC magnetic field (H_{DC}) values in their work, producing a family of MAMMAS curves. These responses were mainly associated with changes in the magnetic moment field derivative and then with the differential susceptibility ($\chi = \frac{\partial M}{\partial H}$), where the peak feature was mainly due to the magnetoresistive effect in magnetite.

2.2 Nickel-zinc ferrites $(Ni_{1-x},Zn_x)Fe_2O_4$

Nickel-zinc ferrite (NZFO: $(Ni_{1-x},Zn_x)Fe_2O_4$) is a solid-state solution whose final members are the nickel ferrite (NFO), a ferrimagnetic inverse spinel ($\downarrow Fe^{3+}$) $[\uparrow Fe^{3+} \uparrow Ni^{2+}]O^{2-}{}_4$; and the zinc ferrite (ZFO), which is a direct spinel with an antiferromagnetic order at low temperatures (Zn^{2+}) $[\uparrow Fe^{3+} \downarrow Fe^{3+}]O^{2-}{}_4$ [1, 44, 45]. The resulting mixed spinel ferrites have a complex cation distribution represented by $(Fe^{3+}{}_{1-x}\,Zn^{2+}{}_x)\ [Fe^{3+}{}_{1+x}Ni^{2+}{}_{1-x}]O^{2-}{}_4$. Having a higher electrical resistivity ($10^6 \Omega{\cdot}m$), NZFO is applied in electronic devices at frequencies between 1 and 500 MHz [2].

The magnetic properties of NZFO can be finely tuned by changing the Zn^{2+} content, as shown in **Figure 11**. Being diamagnetic, when Zn^{2+}content increases, the magnetic moment in the A-sites is diluted. Also, the A-B superexchange coupling interaction between magnetic ions in A- and B-sites is weakened with increasing x, and it causes a non-collinear magnetic ordering. This magnetic order type is known as Yaffet-Kittel (YK), which, as a first approximation, can be regarded as a triangular ordering of magnetic moments between one A-site and two adjacent B-sites [46]. The YK angle increases with x until it reaches 90° for ZFO. So, in magnetization experiments for NZFO, an initial increase up x = 0.4–0.5, was observed while M_s decreases for larger Zn content. On the other hand, the absolute value of the first anisotropy constant decays continuously with x because Ni^{2+} is the main source of crystalline anisotropy in this system. Finally, NZFO with $x > 0.8$ are paramagnetic at room temperature.

Figure 12 shows the extended field microwave absorption (EFMA) measurement. LFA signals more FMR spectrum for bulk powders (**Figure 12a**) and nanocrystalline thin film (**Figure 12b**) in NZFO samples with x = 0.65 [47]. The presence of the FMR signal at high fields and hysteresis around zero applied field due to LFMA are observed. For the bulk powders and thin film of NZFO, the resonance fields are $H_r = 2320\text{Oe}$ and $H_r = 2480\text{Oe}$, respectively, where this difference may be attributed to the nanocrystalline microstructure of the film. Also, the larger FMR linewidth can be due to the stress induced by the interphase between the ferrite and substrate in the thin film.

FMR, LFMA, and MAMMAS studies have also been reported for sintered NZFO samples with x = 0.65 [27, 28]. The microwave absorption dynamics in these samples were interpreted in terms of the temperature dependence of the YK angle at low temperatures [27], while on heating, the changes were due to the Curie transition at T_C = 430 K [28].

In a polycrystalline NZFO sample, the FMR spectrum showed the ferrimagnetic to paramagnetic transition at T_C = 430 K, where the shift in resonance field toward the free electron value is because the internal field vanishes, see **Figure 13a**. Also, ΔH_{pp} decreased with the temperature, **Figure 13b**, where a kink in the linewidth versus temperature curve was observed at T_C and an inflection point at T_Y = 240 K.

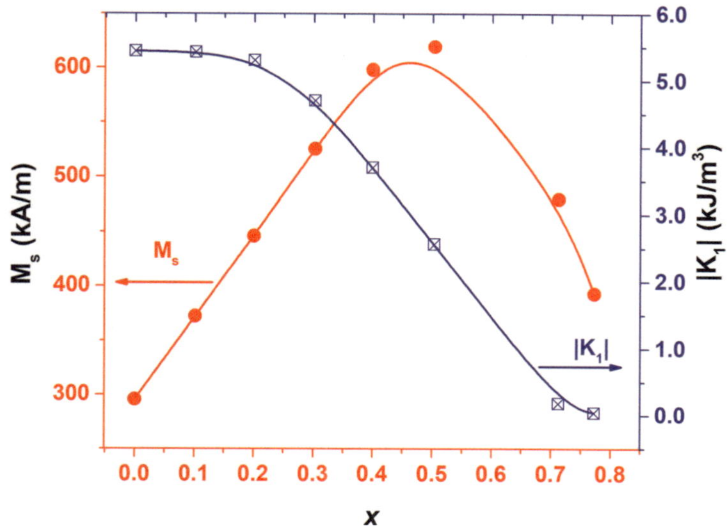

Figure 11.
The variation of the saturated magnetization (M_s– Red circles) and the absolute value of the magnetocrystalline anisotropy constant (K_1.- blue squares) for NZFO as a function of the zinc content (x). With data from [1].

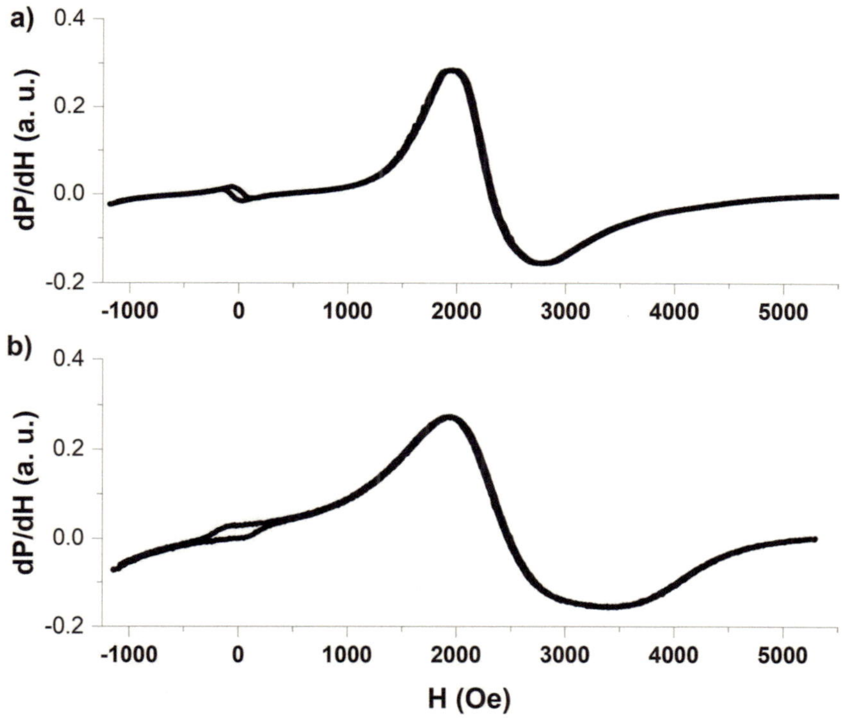

Figure 12.
The extended field microwave absorption in NZFO with x = 0.65 for (a) bulk powders and (b) thin film samples, where LFA signal is evident around H = 0 Oe. Adapted from [47].

On cooling the polycrystalline NZFO sample, MAMMAS response decreased, with a minimum at T_m = 240 K, followed by an increase to low temperature. By correlating the microwave absorption on NZFO with the energy changes due to YK ordering, it was possible to reproduce the main features of the MAMMAS spectrum [48]. The changes in magnetic permeability produce the MAMMAS signal due to the thermal evolution of exchange coupling in the NZFO sample, see **Figure 14**.

On the other hand, the LFA spectrum showed the development of a second component with an opposed phase at $T < 239$ K, see **Figure 15** (up). The onset of this second mode in the LFA spectrum coincides quite well with the $T_m = 240$ K value observed in MAMMAS measurements. Finally, on heating, the LFMA spectrum vanishes for $T > T_C = 430$ K, see **Figure 15** (down).

The study of microwave absorption processes at low magnetic fields is also crucial for developing some applications of soft ferrites for those applications that are expected to operate magnetic devices under small or zero applied fields. Lutsev and Shutkevich [49] reported broadband FMR measurements at 50 MHz-4.0 GHz frequencies in Mn-Zn and Ni-Zn ferrite nanocomposite films. Under these conditions, FMR absorption appears at $H_{DC} < 500$ G, and a significant superposition of LFA and FMR signals is expected. By manipulating the initial demagnetized state with a

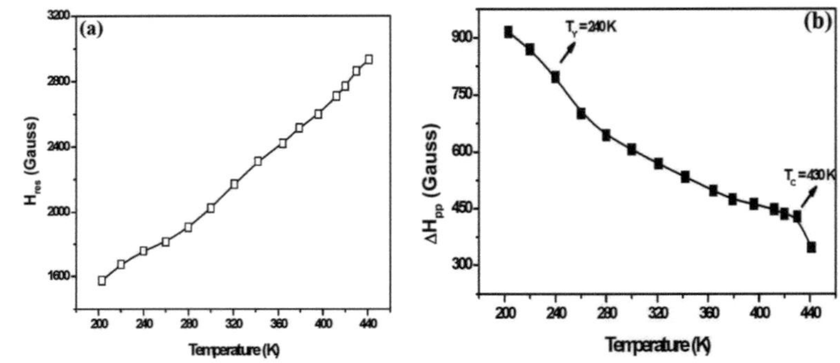

Figure 13.
Temperature dependence of (a) H_{res} *and (b)* ΔH_{pp} *in the 200–440 K temperature range for the polycrystalline NZFO sample. Adapted from [20].*

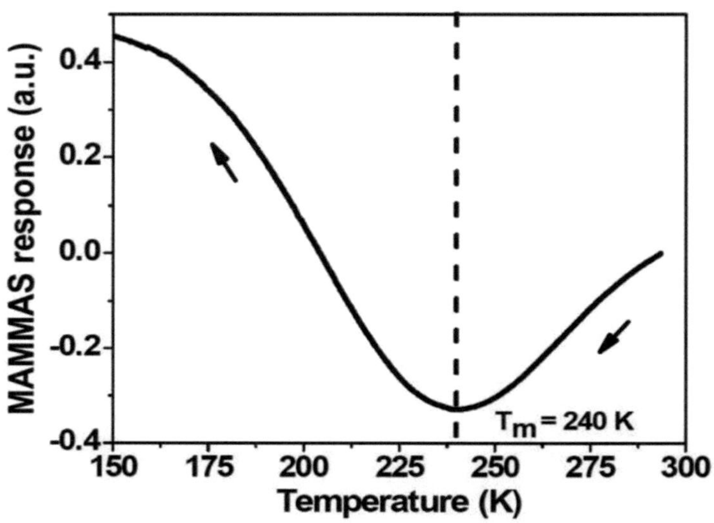

Figure 14.
MAMMAS response for the polycrystalline NZFO sample. Adapted from [20].

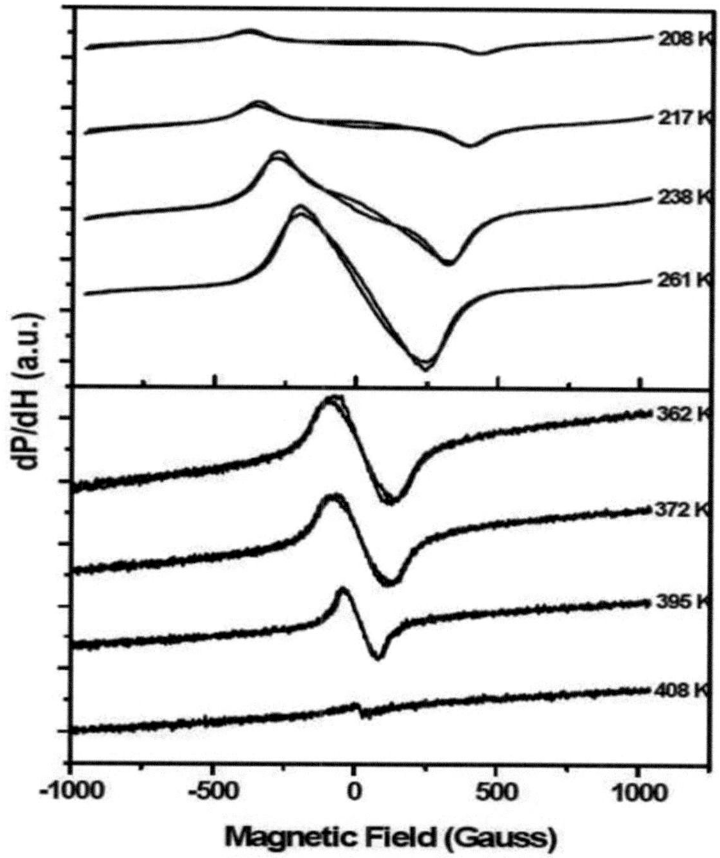

Figure 15.
LFA spectra of the polycrystalline NZFO sample in the 362–408 K (down) and 208–261 K (up) temperature ranges. Adapted from [20].

triangular ac field of low frequency (1 Hz), it was possible to change the absorption levels of the ferrite nanocomposites. This effect was due to the narrowing of the distribution of nanoparticle orientation due to the action of the triangular ac field.

2.3 Yttrium iron garnet (YIG)

YIG ($Y_3Fe_5O_{12}$) is a synthetic magnetic oxide that was independently discovered by Bertaut and Forrat [50], Pauthenet [51], and Geller and Gilleo [52] between 1956 and 1957. YIG adopts the garnet crystal structure, which belongs to the high symmetry space group $Ia\overline{3}d$ (No. 230). Its large unit cell ($a_0 = 12.376$Å) is described as a bcc arrangement of three kinds of polyhedral units, named octahedral (16a), tetrahedral (24d), and dodecahedral (24c), as represented in **Figure 16**. While the large dodecahedral site accommodates the Y^{3+} ion, the ferric ions distribute between octahedral and tetrahedral sites in a 2:3 ratio.

YIG has remarkable magnetic, electric, and magnetic optical properties, which are the basis for its applications in microwave devices for radar and telecommunications [5]. Also, YIG is a soft magnetic material with a low magnetocrystalline anisotropy with cubic anisotropy constants $K_1 = -610$ J/m^3 and $K_2 = -26$ J/m^3 at 300 K [53]. This

Figure 16.
Schematic representation of the crystal structure for yttrium iron garnet (YIG), where it is shown the coordination polyhedral for octahedral (green), tetrahedral (magenta), and dodecahedral (yellow) sites.

behavior is due to the ferric ions, which have a spectroscopic term ^6S meaning that they have a minimum spin-lattice coupling. YIG is also a good insulator with $\rho \approx 10^{10} \Omega \bullet \text{m}$ at 300 K [54] because it only contains trivalent cations, avoiding electrical conduction via an electron hopping mechanism. Its small anisotropy, along with the very high electrical resistivity, gives single crystal YIG the lowest known Gilbert damping factor, being as low as $\alpha_G = 1.3 \times 10^{-4}$; this manifest with the smallest know FMR linewidth (FWHM) of $\Delta H = 0.52$ Oe (0.052 mT) reported by LeCraw (**Figure 17**) [55].

Since their discovery, YIG has been one of the most widely studied materials under the FMR technique. Experiments for measuring its linewidth and damping [56, 57], the power saturation effects, the excitation of spin waves [5, 58–60], and spin pumping [61], and spin-torque transference [62] have been reported.

Although the single crystals and the micrometer-thick epitaxial films have smaller values of linewidth, this parameter is extremely sensitive to microstructure, pores, and magnetic impurities. YIG has been used as a model for studying the FMR relaxation mechanisms in magnetic materials. The difference in FMR properties between single crystals and nanocrystalline YIG materials can be huge, while epitaxial thin films growth onto GGG (100) substrates can exhibit linewidth and damping with low values as $\Delta H = 1.2 - 2.0$ Oe and $\alpha_G = 2.2 \times 10^{-4}$ [63]; the nanocrystalline films growth onto Si (100) can have linewidths as large as $\Delta H = 100 - 400$ Oe, corresponding to Gilbert damping factors of about $\alpha_G \sim 2 \times 10^{-2} - 1 \times 10^{-1}$ [64].

LFA signal in YIG samples has a much lower amplitude, and its linewidth is smaller than in the FMR spectra ($\Delta H_{pp}^{LFA} < \Delta H_{pp}^{FMR}$). The recording of LFA signals in YIG samples usually needs higher powers and gain levels. That is the consequence of the weaker magnetocrystalline anisotropy of YIG as compared with spinel ferrites, so the initial magnetization processes involved in LFA hysteresis play a minor role in the softer garnet ferrite.

A comparative study of the microwave absorption between micropowders and nanopowders in the YIG samples was presented by Sanchez MH et al. [65], where

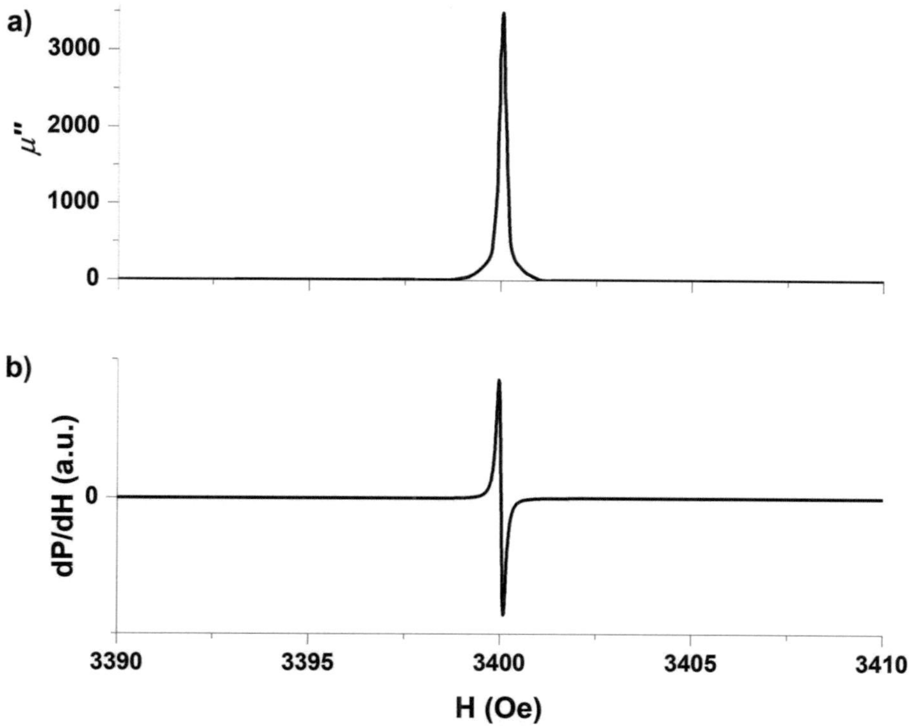

Figure 17.
FMR spectrum for a single crystal YIG sphere measured at 9.3 GHz with H_{DC} applied parallel to the [100] axis. (a) Imaginary permeability representation with data interpolated from LeCraw [55]. (b) Numerically calculated field derivative. The linewidth in (a) is $\Delta H = 520$ mOe, equivalent to $\Delta H_{pp} = 130$ mOe in the dP/dH representation.

the representative FMR spectra are shown in **Figure 18**. The nanopowders in YIG samples show evidence of superparamagnetic (SPM) behavior, which is evidenced by vibrant sample magnetometry (VSM); also, by their nearly constant resonance field because it does not change with the temperature, and it is centered at H_r = 3330 Gauss.

In contrast, the micropowders in the YIG particles exhibit magnetization curves with hysteresis, and a much more complex thermal dependence of their FMR parameters, as expected for magnetic ordering. H_r in nanopowders increases with temperature due to thermal energy weakening of the internal field in the YIG sample. Also, the linewidth decreases with temperature because of magnetocrystalline anisotropy and long-range dipolar interactions, which are proportional to M_s.

As stated before, the LFA signal is a signature of magnetic order in the materials, as we show by the contrast between the FMR and LFA absorption spectra in YIG particles of micrometric and nanometric dimensions [65]. When both LFMA signals are compared, see the insets in **Figure 18a** (up) and **18b** (down) found that the magnetically ordered micrometer-sized YIG particles have an LFA signal with evident hysteresis around zero magnetic fields. Conversely, YIG nanoparticles only exhibited a linear LFMA response with no hysteresis and just a constant slope, which slowly decreases with temperature in coherence with their SPM nature.

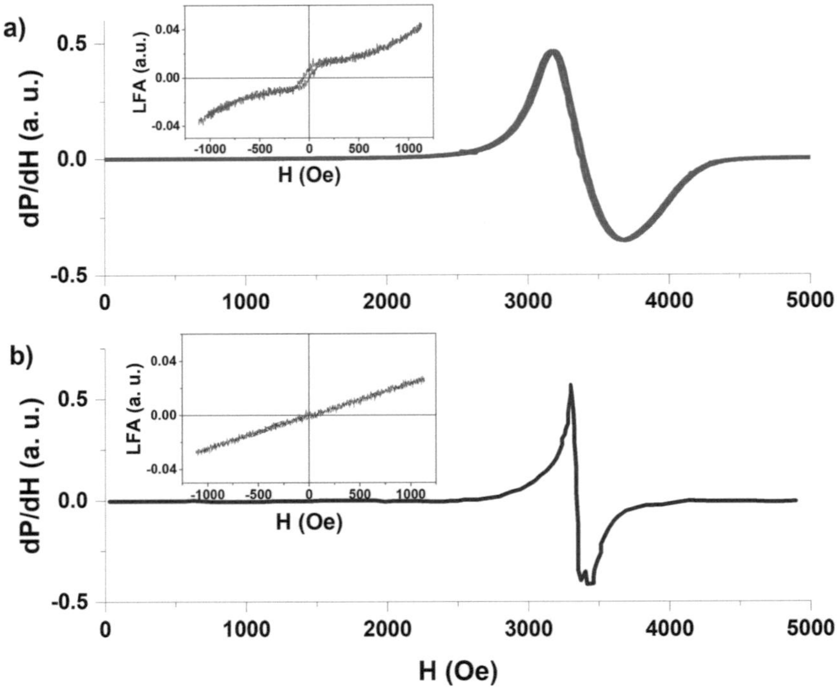

Figure 18.
FMR spectra for (a) micro powders and (b) nanopowder of the YIG samples at 300 K, and their respective LFA signals are shown in the insets. Adapted from [65].

3. Conclusions

Microwave power absorption techniques like FMR, LFA, and MAMMAS are useful for characterizing magnetic ordering, couplings, and anisotropies in soft magnetic ferrites.

In magnetically ordered materials, the presence of the LFA signal is considered a signature of the onset of magnetic ordering. The magnetization processes in unsaturated magnets produce hysteresis on the LFA spectrum. So LFA measurements are useful for detecting magnetic phase transitions and following magnetic anisotropies.

MAMMAS signal is originated from the magnetic field dependence of conduction and magnetization process, while the electric polarization process plays a minor role whenever a cylindrical microwave cavity is implemented.

Acknowledgements

DGAPA-UNAM financially supported this work through the grant **PAPIIT No. IN106621**, and J.F.Barrón thanks DGAPA-UNAM for the postdoc grant.

Conflict of interest

The authors declare no conflict of interest.

Author details

Herlinda Montiel[1*], José Francisco Barrón-López[1] and Guillermo Alvarez[2]

1 Instituto de Ciencias Aplicadas y Tecnología, Universidad Nacional Autónoma de México, Ciudad de México, México

2 Universidad Autónoma de la Ciudad de México

*Address all correspondence to: herlinda.montiel@icat.unam.mx

IntechOpen

© 2023 The Author(s). Licensee IntechOpen. This chapter is distributed under the terms of the Creative Commons Attribution License (http://creativecommons.org/licenses/by/3.0), which permits unrestricted use, distribution, and reproduction in any medium, provided the original work is properly cited.

References

[1] Valenzuela R. Magnetic Ceramics [Internet]. Cambridge: Cambridge University Press; 1994 [cited 2016 May 3]. Available from: http://ebooks.cambridge.org/ref/id/CBO9780511600296

[2] Mattei JL, Chevalier A, Laur V. Ferrite ceramics at microwave Frequencies: Applications and characterization. In: Encyclopedia of Materials: Technical Ceramics and Glasses [Internet]. Amsterdam (NL): Elsevier; 2021. [cited 2023 Jun 16]. pp. 183-205. Available from: https://linkinghub.elsevier.com/retrieve/pii/B9780128035818117655

[3] Hao A, Ning X. Recent advances in spinel ferrite-based thin films: Synthesis, performances, applications, and beyond. Frontiers in Materials. [Internet]. 2021;**8**(November):1-9. DOI: 10.3389/fmats.2021.718869

[4] Mallmann EJJ, Sombra ASB, Goes JC, Fechine PBA. Yttrium iron garnet: Properties and applications review. Solid State Phenomena [Internet]. 2013;**202**: 65-96. Available from: https://www.scientific.net/SSP.202.65

[5] Serga AA, Chumak AV, Hillebrands B. YIG magnonics. Journal of Physics D: Applied Physics [Internet]. 2010;**43**(26):264002. Available from: https://iopscience.iop.org/article/10.1088/0022-3727/43/26/264002

[6] Harris VG, Geiler A, Chen Y, Yoon SD, Wu M, Yang A, et al. Recent advances in processing and applications of microwave ferrites. Journal of Magnetism and Magnetic Materials. 2009;**321**(14):2035-2047

[7] Pardavi-Horvath M. Microwave applications of soft ferrites. Journal of Magnetism and Magnetic Materials. 2000;**215**:171-183

[8] Houbi A, Aldashevich ZA, Atassi Y, Bagasharova Telmanovna Z, Saule M, Kubanych K. Microwave absorbing properties of ferrites and their composites: A review. Journal of Magnetism and Magnetic Materials. North-Holland; [Internet]. 2021;**529**: 167839. Available from: https://linkinghub.elsevier.com/retrieve/pii/S0304885321001153

[9] Vonsovskii SV. Ferromagnetic resonance: The phenomenon of resonant absorption of a high-frequency magnetic field in ferromagnetic substances. In: Vonsovskii SV, editor. Ferromagnetic Resonance. 1st ed. Oxford, UK: Elsevier; [Internet]; 1966. p. 326. Available from: www.sciencedirect.com/science/book/9780080110271

[10] Yalcin O. Ferromagnetic resonance. In: Yalçın O, editor. Ferromagnetic Resonance - Theory and Applications [Internet]. Rijeka (HR): InTech; 2013. pp. 1-46. Available from: http://www.intechopen.com/books/ferromagnetic-resonance-theory-and-applications/ferromagnetic-resonance

[11] Valenzuela R, Alvarez G, Montiel H, Gutiérrez MP, Mata-Zamora ME, Barrón F, et al. Characterization of magnetic materials by low-field microwave absorption techniques. Journal of Magnetism and Magnetic Materials [Internet]. 2008;**320**(14): 1961-1965. Available from: https://linkinghub.elsevier.com/retrieve/pii/S0304885308000565

[12] Alvarez G, Zamorano R. Characteristics of the magnetosensitive nonresonant power absorption of the microwave by magnetic materials.

Journal of Alloys and Compounds. 2004; **369**(1–2):231-234. Available from: https://linkinghub.elsevier.com/retrieve/pii/S0925838803009630

[13] Rado GT, Wright RW, Emerson WH, Ferromagnetism at Very High Frequencies. III. Two mechanisms of dispersion in a ferrite. Physics Review [Internet]. 1950;**80**(2):273-280. DOI: 10.1103/PhysRev.80.273

[14] Landau LD, Lifshits EM. Continuum Electrodynamics. 2nd ed. Vol. 8. Theoretical physics: Butterworth-Heinemann; 1982

[15] Wampler J, Hua N, Kukreja R, Ramìrez JG, Basaran AC, Fullerton EE, et al. Electromagnetic origin of the microwave absorption response of Fe3O4 thin films. Physical Review B. 2022;**106**(6):L060402. DOI: 10.1103/PhysRevB.106.L060402

[16] Maksymov IS, Kostylev M. Broadband stripline ferromagnetic resonance spectroscopy of ferromagnetic films, multilayers and nanostructures. Physica E: Low-dimensional Systems and Nanostructures. [Internet]. 2015;**69**: 253-293. DOI: 10.1016/j.physe.2014.12.027

[17] Lo CK. Instrumentation for ferromagnetic resonance spectrometer. In: Yalcin O, editor. Ferromagnetic Resonance - Theory and Applications. [Internet]. Rijeka: InTech; 2013. pp. 47-62. DOI: 10.5772/56069

[18] Heinrich B. Radio Frequency Techniques. In: Heinrich B, JAC B, editors. Ultrathin Magnetic Structures II [Internet]. Berlin, Heidelberg: Springer Berlin Heidelberg; 1994. pp. 195-296. DOI: 10.1007/3-540-27166-X_3

[19] Webb A. Cavity- and waveguide-resonators in electron paramagnetic resonance, nuclear magnetic resonance, and magnetic resonance imaging. Progress in Nuclear Magnetic Resonance Spectroscopy [Internet]. 2014;**83**:1-20. DOI: 10.1016/j.pnmrs.2014.09.003

[20] Montiel H, Alvarez G. Detection of magnetic transitions by means of ferromagnetic resonance and microwave absorption techniques. In: Yalcin O, editor. Ferromagnetic Resonance - Theory and Applications [Internet]. Rijeka: InTech; 2013. pp. 63-92. DOI: 10.5772/55962

[21] Poole CP. Electron Spin Resonance: A Comprehensive Treatise on Experimental Techniques. 2nd ed. New York: John Wiley & Sons; 1983

[22] Gurevich AG, Melkov GA. Magnetization Oscillations and Waves [Internet]. 1st ed. Boca Raton (FL): CRC Press; 1996. Available from: https://www.crcpress.com/Magnetization-Oscillations-and-Waves/Gurevich-Melkov/p/book/9780849394607

[23] Gilbert TL. Classics in magnetics a phenomenological theory of damping in ferromagnetic materials. IEEE Transactions on Magnetics. [Internet]. 2004;**40**(6):3443-3449. DOI: 10.1109/TMAG.2004.836740

[24] Gutiérrez MP, Alvarez G, Montiel H, Zamorano R, Valenzuela R. Study of the Verwey transition in magnetite by low field and magnetically modulated nonresonant microwave absorption. Journal of Magnetism and Magnetic Materials [Internet]. 2007;**316**(2):e738-e740. Available from: https://linkinghub.elsevier.com/retrieve/pii/S0304885307004155

[25] Nabereznykh VP, Tsindlekht MI. Nonresonant microwave absorption in

nickel near the curie point. Journal of Experimental and Theoretical Physics Letters [Internet]. 1982;**36**(4):157. Available from: http://jetpletters.ru/ps/1332/article_20129.shtml

[26] Owens FJ. DC magnetic field-dependent microwave absorption in CMR material, La0.7Sr0.3MnO3. Journal of Physics and Chemistry of Solids [Internet]. 1997;**58**(9):1311-1314. Available from: https://www.sciencedirect.com/science/article/pii/S0022369797000401

[27] Alvarez G, Montiel H, Barron JF, Gutierrez MP, Zamorano R. Yafet–Kittel-type magnetic ordering in Ni0.35Zn0.65Fe2O4 ferrite detected by magnetosensitive microwave absorption measurements. Journal of Magnetism and Magnetic Materials [Internet]. 2010;**322**(3):348-352. Available from: https://linkinghub.elsevier.com/retrieve/pii/S0304885309009536

[28] Montiel H, Alvarez G, Gutiérrez MP, Zamorano R, Valenzuela R. Microwave absorption in Ni-Zn ferrites through the curie transition. Journal of Alloys and Compounds. 2004:**369**(1-2):141-143. DOI: 10.1016/j.jallcom.2003.09.074

[29] Lee SJ, Tsai CC, Cho H, Seo M, Eom T, Nam W, et al. Hysteretic characteristics of low-field microwave absorption of a Co thin film. Journal of Applied Physics [Internet]. 2009; **106**(6):063922. DOI: 10.1063/1.3204648

[30] Montiel H, Alvarez G, Betancourt I, Zamorano R, Valenzuela R. Correlations between low-field microwave absorption and magnetoimpedance in Co-based amorphous ribbons. Applied Physics Letters [Internet]. 2005;**86**(7):1-3. DOI: 10.1063/1.1861959

[31] Cervellino A, Frison R, Cernuto G, Guagliardi A, Masciocchi N. Lattice parameters and site occupancy factors of magnetite–maghemite core–shell nanoparticles. A critical study. Journal of Applied Crystallography [Internet]. 2014;**47**(5):1755-1761. DOI: 10.1107/S1600576714019840

[32] Walz F. The Verwey transition - a topical review. Journal of Physics. Condensed Matter [Internet]. 2002; **14**(12):R285-R340. Available from: https://iopscience.iop.org/article/10.1088/0953-8984/14/12/203

[33] Muxworthy AR, McClelland E. Review of the low-temperature magnetic properties of magnetite from a rock magnetic perspective. Geophysical Journal International [Internet]. 2000; **140**(1):101-114. Available from: https://academic.oup.com/gji/article/140/1/101/707798

[34] Rozenberg GK, Pasternak MP, Xu WM, Amiel Y, Hanfland M, Amboage M, et al. Origin of the Verwey transition in magnetite. Physical Review Letters [Internet]. 2006;**96**(4):045705. Available from: https://journals.aps.org/prl/abstract/10.1103/PhysRevLett.96.045705

[35] Senn MS, Wright JP, Attfield JP. Charge order and three-site distortions in the Verwey structure of magnetite. Nature [Internet]. 2012;**481**(7380):173-176. Available from: https://www.nature.com/articles/nature10704

[36] Senn MS, Wright JP, Cumby J, Attfield JP. Charge localization in the Verwey structure of magnetite. Physical Review B [Internet]. 2015;**92**(2):024104. DOI: 10.1103/PhysRevB.92.024104

[37] Bohra M, Agarwal N, Singh V. A short review on Verwey transition in nanostructured Fe3O4 materials

[Internet]. Journal of Nanomaterials. 2019;**2019**:1-18. Available from: https://www.hindawi.com/journals/jnm/2019/8457383/

[38] Gridin VV, Hearne GR, Honig JM. Magnetoresistance extremum at the first-order Verwey transition in magnetite (Fe3O4). Physical Review B [Internet]. 1996;**53**(23): 15518-15521. DOI: 10.1103/PhysRevB.53.15518

[39] Gasparov LV, Tanner DB, Romero DB, Berger H, Margaritondo G, Forró L. Infrared and Raman studies of the Verwey transition in magnetite. Physical Review B. 2000;**62**(12): 7939-7944. DOI: 10.1103/PhysRevB.62.7939

[40] Alvarez G, Montiel H. Microwave absorption dynamics in Fe3O4 nanopowders around Verwey transition. Journal of Physics Conference Series [Internet]. 2022;**2307**(1):012040. Available from: https://iopscience.iop.org/article/10.1088/1742-6596/2307/1/012040

[41] Bickford LR. Ferromagnetic resonance absorption in magnetite single crystals. Physics Review [Internet]. 1950;**78**(4):449-457. Available from: https://journals.aps.org/pr/abstract/10.1103/PhysRev.78.449

[42] Srivastava A, Singh AV, Mohammadi JB, Mewes C, Gupta A, Mewes T. Ferromagnetic resonance study of the Verwey phase transition of magnetite thin film on $MgGa_2O_4$(001) substrate. IEEE Transactions on Magnetics [Internet]. 2020;**56**(12):1-6. Available from: https://ieeexplore.ieee.org/document/9201012/

[43] Schlömann E. Ferromagnetic resonance in polycrystalline ferrites with large anisotropy-I. general theory and application to cubic materials with a negative anisotropy constant. Journal of Physics and Chemistry of Solids [Internet]. 1958;**6**(2–3):257-266. Available from: https://www.sciencedirect.com/science/article/pii/0022369758901021

[44] Kaur A, Sharma P, Bhardwaj S, Kumar M, Sharma I, Batoo KM, et al. A review on synthesis and characterizations of mixed nickel-zinc ferrites. In: Sharma G, Kumar A, Dhiman P, editors. Ferrite - Nanostructures with Tunable Properties and Diverse Applications. Vol. 1112. Materials Research Foundations. Millersville (PA): Materials Research Forum LLC; 2021. pp. 189-217

[45] Dionne GF. Magnetic oxides [Internet]. In: Magnetic Oxides. Boston, MA: Springer US; 2009. pp. 1-466. Available from: http://link.springer.com/10.1007/978-1-4419-0054-8

[46] Yafet Y, Kittel C. Antiferromagnetic arrangements in ferrites. Physics Review [Internet]. 1952;**87**(2):290-294. Available from: https://journals.aps.org/pr/abstract/10.1103/PhysRev.87.290

[47] Barron-Lopez JF, Montiel H, Mata-Zamora ME, Alvarez G. Obtención y caracterización de películas magnéticas de ferrita de Ni-Zn por depósito electroforético. Superf y Vacio. 2011; **24**(4):126-131. Available from: http://ref.scielo.org/hdhpk7

[48] Murthy NSS, Natera MG, Youssef SI, Begum RJ, Srivastava CM. Yafet-kittel angles in zinc-nickel ferrites. Physics Review [Internet]. 1969; **181**(2):969-977. DOI: 10.1103/PhysRev.181.969

[49] Lutsev L, Shutkevich V. Sharp increase of microwave absorption in

nonequilibrium MnZn- and NiZn-nano ferrites. Journal of Physics D: Applied Physics [Internet]. 2016;**49**(50):505002. Available from: https://iopscience.iop.org/article/10.1088/0022-3727/49/50/505002

[50] Bertaut F, Forrat F. Structure des ferrites ferrimagnetiques des terres rares. Comptes Rendus hebdomadaires des séances de l'Académie des Sciences. 956;**242**(1):382

[51] Pauthenet R. Spontaneous magnetization of some garnet ferrites and the Aluminum substituted garnet ferrites. Journal of Applied Physics. 1958; **29**(3):253. Available from: http://scitation.aip.org/content/aip/journal/jap/29/3/10.1063/1.1723094

[52] Geller S, Gilleo MA. Structure and ferrimagnetism of yttrium and rare-earth–iron garnets. Acta Crystallographica. 1957;**10**(3):239-239. Available from: http://scripts.iucr.org/cgi-bin/paper?S0365110X57000729

[53] Stancil D, Prabhakar A. Spin Waves. Boston, MA: Springer US; [Internet]; 2009. DOI: 10.1007/978-0-387-77865-5

[54] Larsen PK, Metselaar R. Electrical properties of yttrium iron garnet. In: Paoletti A, Societa Italiana di Fisica, editors. Physics of Magnetic Garnets: Proceedings of the International School of Physics (Enrico Fermi) [Internet]. Course LXX; 27 June-9 July 1977. Amsterdam (NL): North-Holland Publishing Company; 1978. pp. 417–444. Available from: https://pure.tue.nl/ws/files/1871955/620146.pdf

[55] LeCraw RC, Spencer EG, Porter CS. Ferromagnetic resonance line width in yttrium iron garnet single crystals. Physics Review. 1958;**110**(6):1311-1313. DOI: 10.1103/PhysRev.110.1311

[56] Glass HL, Elliot MT. Attainment of the intrinsic FMR linewidth in yttrium iron garnet films grown by liquid phase epitaxy. Journal of Crystal Growth. 1976; **34**(2):285-288

[57] Stognij A, Lutsev L, Novitskii N, Bespalov A, Golikova O, Ketsko V, et al. Synthesis, magnetic properties, and spin-wave propagation in thin Y3Fe5O12 films sputtered on GaN-based substrates. J Phys D Appl Phys [Internet]. 2015;**48** (48):485002. Available from: https://iopscience.iop.org/article/10.1088/0022-3727/48/48/485002

[58] Collet M, Gladii O, Evelt M, Bessonov V, Soumah L, Bortolotti P, et al. Spin-wave propagation in ultra-thin YIG-based waveguides. Applied Physics Letters. 2017;**110**:092408

[59] Yu H, d'Allivy KO, Cros V, Bernard R, Bortolotti P, Anane A, et al. Magnetic thin-film insulator with ultra-low spin wave damping for coherent nano magnonics. Scientific Reports. 2014;**4**:6848. Available from: https://www.nature.com/articles/srep06848

[60] Rezende SM. Fundamentals of Magnonics. Cham, Switzerland: Springer; 2020

[61] Sun Y, Chang H, Kabatek M, Song YY, Wang Z, Jantz M, et al. Damping in yttrium iron garnet nanoscale films capped by platinum. Physical Review Letters. 2013;**111**(10): 106601. Available from: http://link.aps.org/doi/10.1103/PhysRevLett.111.106601

[62] Klinger S, Amin V, Geprägs S, Ganzhorn K, Maier-Flaig H, Althammer M, et al. Spin-torque excitation of perpendicular standing spin

waves in coupled YIG=Co Heterostructures. Physical Review Letters. 2018;**120**:127201

[63] Onbasli MC, Kehlberger A, Kim DH, Jakob G, Kläui M, Chumak AV, et al. Pulsed laser deposition of epitaxial yttrium iron garnet films with low Gilbert damping and bulk-like magnetization. APL Materials [Internet]. 2014;**2**(10):106102. DOI: 10.1063/1.4896936

[64] Yamada K, Kogiso K, Shiota Y, Yamamoto M, Yamaguchi A, Moriyama T, et al. Dependence of Gilbert damping constant on microstructure in nanocrystalline YIG coatings prepared by co-precipitation and spin-coating on a Si substrate. Journal of Magnetism and Magnetic Materials. 2020;**513**:167253. DOI: 10.1016/j.jmmm.2020.167253

[65] Montiel H, Alvarez G, Conde-Gallardo A, Zamorano R. Effect of the particle size on the microwave absorption in the yttrium-iron garnet. Journal of Nano Research. 2014;**28**:73-81. Available from: https://www.scientific.net/JNanoR.28.73

Chapter 6

Comparison of Magnetic and Electrical Properties of Manganese-Doped Cobalt Ferrite Nanoparticles

Md. Ziaul Ahsan

Abstract

Cobalt ferrites have been regarded as one of the competitive candidates in diversified applications because of their tailored electrical and magnetic properties by tuning structural parameters with doping of transition and/or rare earth elements. To meet the ever-growing technological demands, varieties of new compositions of cobalt ferrites with diverse properties are continuously developed and produced through different synthetic routes across the globe and are made available to the consumers. In the recent past, three compositions of manganese-doped cobalt ferrite nanoparticles have been developed through a solid-state reaction route by using the planetary ball milling technique. In this chapter, a comparison on their novelty in respect of their magnetic and electrical properties is made to spot their possible applications. Besides, few suggestions have been put forwards for further investigations to meet the challenges in the days to come.

Keywords: permeability, magnetic modulus, permittivity, electric modulus, resistivity, impedance spectroscopy

1. Introduction

Cobalt ferrite ($CoFe_2O_4$) is a ceramic metal oxide having an inverse spinel structure with the basic composition formula $(B)^{Tet}$ $[AB]^{Oct}O_4$. Its tetrahedral and octahedral sites are formed by oxygen ions in a face-centered cubic crystal structure. Fe^{3+} ions occupy the tetrahedral site, and Fe^{3+} and Co^{2+} ions occupy the octahedral site in equal proportionate. This cobalt ferrite exhibits the property of an electrical insulator and magnetic conductor/semiconductor similar to ferrite materials with a large number of diversified applications. An excellent ability to incorporate metals ions in its lattice attracts materials scientists to synthesize and optimizes them with new interesting properties. The manganese-doped cobalt ferrites mostly demonstrate magnetic, electrical, and catalytic properties, which are correlated to their structural properties. The doping of metallic ions having an ionic radius less than unity (<1 Å) changes the structural and morphological properties in the doped cobalt ferrites [1–3]. The cation distribution over the tetrahedral and octahedral sites governs tuning and/or tailoring the structural properties of the doped cobalt ferrites that in turn lead to change their magnetic, electrical, and catalytic behavior. This distribution of cations depends on

IntechOpen

various factors like the size of the dopant ion, charge possessed by the dopant ion, site preference of the dopant ion, synthesis technique, preparation conditions, heat treatments, and the extent of dopant. Besides, the particle size plays a significant role to change the structural, morphological, magnetic, electrical, and catalytic properties of doped cobalt ferrite nanoparticles [4, 5]. Over the years, cobalt ferrites doped with transition (TE) and rarearth (RE) elements have been synthesized by various techniques and optimized for multifarious applications. Recently, nanoparticles of cobalt ferrites doped with metallic ions have received renewed attention from the scientific community across the globe to optimize them in medical science, catalyst, environmental, and sensor applications [6–12]. By the time, a series of investigations on structural, electrical, and magnetic properties of manganese-doped cobalt ferrite nanoparticles were reported with composition formulas $Co_{1-x}Mn_xFe_2O_4$, $CoMn_xFe_2O_4$, and $Co_{1+x}Mn_xFe_{2-x}O_4$, where x refers to the concentration levels, $0.125 \leq x \leq 0.5$. In the first composition, manganese (Mn) was substituted for cobalt (Co), and in the second composition, Mn was substituted for Iron (Fe). Both of them were stoichiometric compositions. In the third composition, Mn substituted for Fe, but the same concentration/amount of Co was added in this composition to make this system a non-stoichiometric one [3, 4, 13–23]. The main purpose of this chapter is to consolidate and make a comparison of the obtained results to focus their possible applications in light of their electrical and magnetic properties and also suggest further investigations. In this chapter, significant structural, magnetic, and electrical properties of manganese-doped cobalt ferrite nanoparticles with stoichiometric and non-stoichiometric compositions have been discussed along with their possible applications considering their academic and research values.

1.1 Theoretical aspects

1.1.1 Solid-state reaction and its mechanism

Solid-state reaction simply refers to all solventless processes leading a solid reactant to a solid product [24]. It occurs within the rigid constraint environment of the crystal lattice. The solid-state reaction provides the extreme case for evaluating the effect of intermolecular forces on a reaction and their influence on reaction mechanism and direction. The confined environment of the reactant crystal lattice can control the kinetic features of a reaction and hence the nature of the products. This solid-state reaction can take place with minimum requirements of energy and atomic or molecular motions according to the topochemical principle, proposed by Cohen and Schmidt. The advantages of solid-state reactions are as follows: (i) the atom economic nature of the solid-state reactions and the limited formation of side products, (ii) no solvents are required in the reaction so no waste disposed of issue associated with the solvent comes out. As such products do not require vigorous purification, (iii) the constrained environment can lead to novel chemical reactions, and (iv) this reaction is a bit faster than others [25]. To understand the mechanism of the solid-state reaction, let us consider the thermal reaction of two crystals of compounds A and B, which are in intimate contact across one face as shown in **Figure 1**. When no melt is formed during the reaction, the reaction has to occur initially at the points of contact between A and B compounds and later by diffusion of the constituents through the product's phase. Hence, there will be two stages in the solid-state reaction, namely:

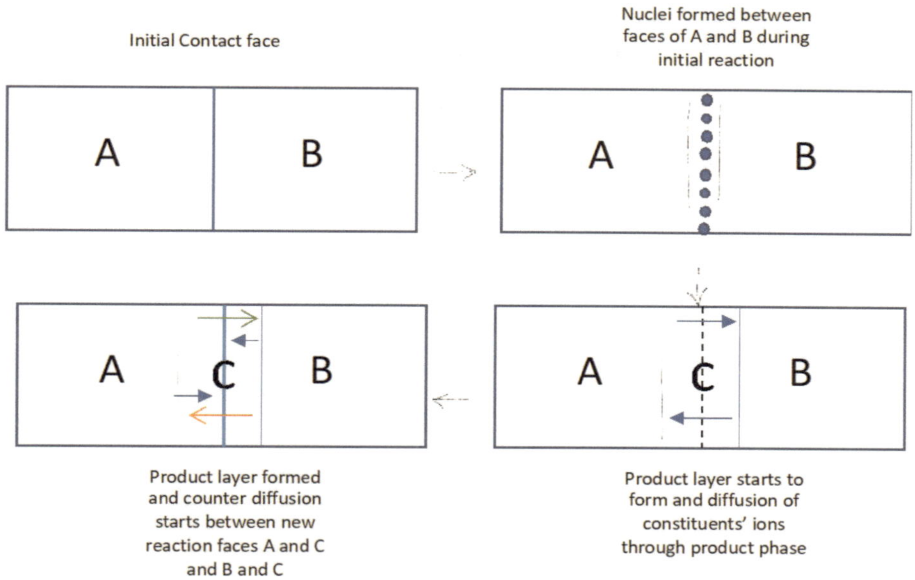

Figure 1.
Schematic diagram of the mechanism of solid-state reaction. (i) The first stage of the reaction is the formation of nuclei of the product phase C at the interface between A and B. After nucleation of product C has occurred, a product layer is formed as shown in b. (ii) At this stage, there are two reaction interfaces, one between A and C and another between C and B. For further reaction to occur, counter-diffusion of ions from A and B must occur through the existing product layer C to the new reaction interfaces.

As the reaction progresses, the product layers become thicker, which result in progressively longer diffusion paths and slower reaction rates because the product layer between the reacting particles acts as barriers. So, the reaction is controlled by the lattice diffusion and the rate law has a parabolic form and is given by $\frac{dx}{dt} = \kappa x^{-1}$, where x is the amount of reaction, which is equal to the thickness of the growing product layer. Ions are normally regarded as being trapped on their appropriate lattice sites, and it is difficult for them to move to the adjacent sites. Only at very high temperatures, the ions have sufficient energy to diffuse through the crystal lattice. As a thumb rule, two-third of the melting temperatures of one constituent is sufficient to achieve diffusion sufficiently and hence to enable the solid-state reaction. From this discussion, it is clear that the reaction between two solids may not occur even if thermodynamic considerations favor the product formation. So, three important factors influence the rate of reaction between solids as (i) the area of contact between reacting solids and hence their surface area, (ii) the rate of nucleation of the product phase, and (iii) the rate of diffusion through the product's phase. However, apart from the problems arising from the nucleation and diffusion, this method suffers from several additional disadvantages such as: (i) the undesirable phase may be formed, (ii) the homogeneous distribution of dopants sometimes is difficult to achieve, (iii) there is a limited possibility to monitor the progress of the reaction, (iv) because of this difficulty, mixtures of reactants and products are frequently obtained, so separating the desired products from the mixture is generally difficult, and (v) in many systems, the reaction temperature cannot be raised as high as necessary for the reasonable reaction, because one or more components of the reacting mixture may be volatile. So, it is necessary to optimize critical parameters. A variety of techniques are there for achieving the products by this solid-state reaction. Despite some disadvantages like the irregular shape of the particle, the possibility to be infected by impurities, ball milling is a versatile and cost-effective technique owing to having control over the particle size to some extent by the milling time [26].

1.1.2 Basics on ferrite structure

The spinel is any of a class of minerals, whose name coined from the mineral spinel $MgAl_2O_3$. The general formula of the spinel is given by $A^{2+}B_2^{3+}O_4^{2-}$ when it is crystallized in the cubic (isometric) crystal systems. In it, the oxide ions (anions) are arranged in a cubic close-packed lattice. ***A*** and ***B*** are the cations bearing charges 2^+ and 3^+ to have it overall charge neutral. The cations ***A*** and ***B*** occupy some or all of the tetrahedral and octahedral sites in the lattice depending on the preferences of sites. The A cations are usually divalent and B cations trivalent. But other combinations are also possible. The anions are usually oxygen; when the other chalcogenides[1] constitute the anion sublattices, then they are referred to as a Thiospinel. It notable here that *A* and *B* can also be the same metal but with different valences. For example, magnetite, whose formula can be written as Fe_3O_4 or $Fe^{2+}Fe_2^{3+}O_4^{2-}$, is the most abundant member of the spinel group. The spinels are grouped in series by the ***B*** cations. The main spinel groups are shown in **Table 1** with respective generic formula, where ***A*** cations are exhibited by ***X*** for convenience to denote its variable constituents.

They usually have *fcc* packing of anions. In the spinel structure, there are 64 tetrahedral and 32 octahedral sites. 1/8th tetrahedral and 1/2th octahedral sites are usually occupied in its unit cell. Accordingly, a spinel unit cell is made up of eight *fcc* cells of oxygen ions in the configuration $2 \times 2 \times 2$. So it is a big structure consisting of 32 oxygen ions, 8 *A* ions, 16 *B* ions, and a total of 56 ions [3]. Thus, a spinel unit cell contains two types of sublattices as depicted in **Figure 2**. These two types of sublattices repeat alternately in a three-dimensional array to produce a spinel unit cell, which ultimately requires eight sublattices.

Group	Cations		Anions	Formula	*X = A cation*	Remark
	A	B				
Aluminum Spinel	*X*	Aluminum (Al)	Oxygen (O)	XAl_2O_4	*X = Mg, Be, Zn, Fe, Mn*	
Iron Spinel	*X*	Iron (Fe)	Oxygen (O)	XFe_2O_4	*X = Co, Cu, Mn, Mg, Fe, Ni, Ti, (Zn, Fe)*	Ferrites
Chromium Spinel	*X*	Chromium (Cr)	Oxygen (O)	XCr_2O_4	*X = Fe, Mg, Zn*	

1. Franklinite is excluded from the given formula because it is represented as (Fe, Mn, Zn)(Fe, Mn)$_2$O$_4$.
2. The heterogeneity of spinel group members varies based on composition with ferrous and magnesium-based members varying greatly as in solid solution, which requires similarly sized cations. However, ferric and aluminum-based spinels are almost entirely homogeneous due to their large size difference.

Table 1.
Main spinel group with respective generic formula.

[1] The elements in group 16 of the periodic table, which is also called oxygen family like sulfur (S), selenium (Se), tellurium (Te) and polonium (Po).

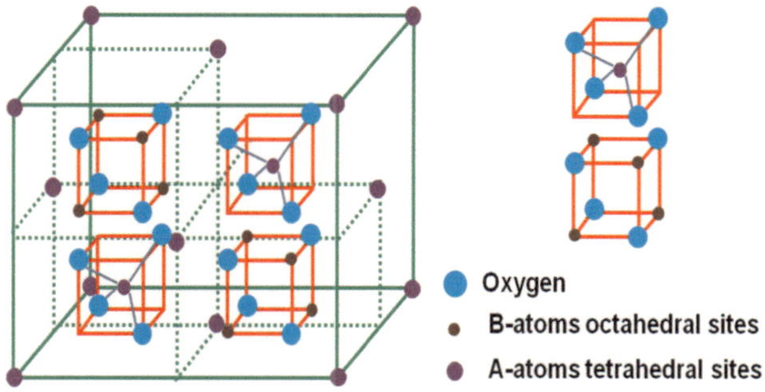

Figure 2.
Schematic diagram of the unit cell structure of spinel ferrite [4, 5].

The interstices available in an ideal close-packed structure of rigid oxygen anions can incorporate in the tetrahedral sites, only the metal ion with a radius $r_{tetra} \leq 0.30$ Å and in octahedral sites, only ions with a radius $r_{oct} \leq 0.55$ Å. To accommodate cations like Co^{2+}, Cu^{2+}, Mg^{2+}, Ni^{2+}, and Zn^{2+}, the lattice has to be expanded. The difference in the expansion of octahedral and tetrahedral sites is characterized by a parameter called oxygen parameter (u). In an ideal spinel, the tetrahedral and octahedral sites are enlarged in the same ratio, and accordingly, the distance between the tetrahedral is (0 0 0) and the oxygen site 3*a* /8 or 0.375 *a* where '*a*' is the lattice constant [3, 6] and hence u_{ideal} = 3/8 Or p.375. The tetrahedral sites are usually too small for the metal ions. Accordingly, the incorporation of divalent metal ions in tetrahedral sites induces a larger expansion of the tetrahedral sites, leading to a large value for 'u' than the ideal value. The tetrahedral sites are expanded by an equal displacement of the four oxygen ions onwards, along the body diagonals of the cube, still occupying the corners of an expanded regular tetrahedron. The four oxygen ions of the octahedral sites are shifted in such a way that this oxygen octahedron shrinks by the same amount as the first expands. As such, a remarkable characteristic of spinel structure is that it can form an extremely wide variety of total solid solutions. This means that the composition of a given ferrite can be strongly modified, while the basic crystalline structure remains the same. Depending on how cations occupy the interstices, spinel structure can be categorized as (i) normal spinel structure (NSS), (ii) inverse spinel structure (ISS), and mixed spinel structure (MSS). The description of all these categories of spinel structure is summarized in **Table 2**.

Structure	Chemical formula	Sites		Remark
		Tetrahedral (t_h) Coordination = 4	Octahedral [O_h] Coordination = 6	
NSS	$(A^{2+})\left[B^{3+}\right]_2O_4^{2-}$	All A^{2+} cations occupy	All B^{3+} cations occupy	
ISS	$(B)[AB]O_4$	½ B^{3+} cations occupy	½ B^{3+}, and all A^{2+} cations occupy	
MSS	$(A_{1-\delta}B_{\delta})\left[A_{\frac{\delta}{2}}B_{1-\frac{\delta}{2}}\right]_2O_4$	1. 'δ' is called the inversion degree. It can take value between 0 and 1, i.e., $0 < \delta < 1$. 2. If $\delta = 0$, then it becomes NSS, and if $\delta = 1$, then it becomes ISS. 3. So, 'δ' can be referred to as a tunable factor between NSS and ISS in MSS.		

Note: cations in NSS: $A^{2+} - \frac{2}{4}$ and $B^{3+} - \frac{3}{6}$; oxygen valance: $\frac{2}{4} \times 1 + \frac{3}{6} \times 3 = 2$.

Table 2.
Summary description of various categories of spinel structure.

Degree of inversion (i)	Occupation of sites	Structure of spinel ferrites	Generic formula
0	A site by $(M_{1-x} - M'_x)$ and B site by Fe	NSS	$(M_{1-x} - M'_x)^A [Fe_2]^B O_4$
1	A site by Fe and B site by $(M_{1-x} - M'_x)$ and Fe	ISS	$(Fe)^A \left[(M_{1-x} - M'_x)Fe\right]^B O_4$
0.5	A site by ½ M and ½ Fe, B site by ½ $(M_{1-x} - M'_x)$ and 3/2 Fe	MSS	$\left\langle (M_{1-x} - M'_x)_{0.5} Fe_{0.5} \right\rangle^A \left[(M_{1-x} - M'_x)_{0.5} Fe_{1.5}\right]^B O_4$

Note: 1. M' *represents the impurity of the same valance. 2.* x *represents the wt% or at%.*

Table 3.
Summary description of doped spinel ferrites' structure.

Doped ferrites are fabricated classes of ferrites where in some impurity atoms are doped in their existing ferrous or ferric matrix. They may be magnetic or ceramic compounds depending upon their compositions and also be of spinel structure or perovskite structure. The possibilities of doping in producing doped ferrites along with the generic formula of such ferrites are in its brief discussion focusing on the spinel structure. In this case, it is possible to introduce impurity of the same valance either in A or B and even in both sites by maintaining wt% or at% of M and impurity (M'). Accordingly, the generic formula for doped spinel ferrites may be written as summarized in **Table 3**.

1.2 Comprehensive results and novelty of Mn-doped cobalt ferrite nanoparticles

1.2.1 Used technique for synthesis and characterization

The samples of three systems of Mn-doped $CoFe_2O_4$ ferrite nanoparticles were prepared through the above-explained solid-state reaction route by using the ball milling technique (Model). Laboratory graded oxide powders of cobalt (Co), manganese (Mn), and iron (Fe) were mixed in a mortar with a pestle for 2 h and then ball milled for 10 h. The ball-milled powders were then calcined at 750°C in a furnace for 1 h in the air atmosphere. The calcined powders were pelletized in the form of disc and toroid for electrical and magnetic measurements, respectively [17, 18, 22]. These toroid and disc-shaped samples were again sintered at 1050°C in the same furnace for 1 h in the air atmosphere. The Waynekerr impedance analyzer 6500B was used to measure electrical and magnetic properties over the frequency band 100 Hz–120 MHz. The calcined powders were used in the EMPYREAN (PAN analytical) XRD machine to record diffraction data for structural analysis and field emission scanning electron microscope (FESEM, JEOL, JSM-7600F) integrated with energy dispersive spectroscopy (EDS) for morphological and elemental analysis. The homemade vibrating sample magnetometer (VSM) was used to measure DC magnetization by using the calcined powders for the purpose [17, 22–24]. The formulas used for calculation of different parameters associated with structural, electric, and magnetic properties to analyze the investigated samples have been presented in tabular form as Appendix A.

1.2.2 Structural properties

The XRD patterns for all three compositions showed the sharp peaks and confirmed their crystallinity with a single-phase spinel structure. They were closely resembled and well-matched with the standards JCPDS card No. 22-1086 for $CoFe_2O_4$ [24]. Their FESEM micrographs (representative micrographs shown in Appendix B) showed little agglomerated particles with nearly spherical shapes and pores [17, 18, 22]. Their EDS spectrum also confirmed the presence of their compositional elements without any impurities. The Xpert pro-High Score Plus software was used to estimate their structural parameters from their strongest peaks correspond to miller plane (311), and their particle size was determined by using the Image-J software and found to be in the nanoscaled range [17, 18, 22]. All these structural and morphological parameters have been reproduced and presented in tabular form in **Table 4** to demonstrate their tunability by the Mn content or concentration level (x) and to analyze the variation nature system to system:

The obtained lattice constant (*a*) was 8.380–8.383 Å for $Co_{1-x}Mn_xFe_2O_4$ composition, 8.358–8.415 Å for $CoMn_xFe_{2-x}O_4$ composition, and 8.4–8.41 Å for $Co_{1+x}Mn_xFe_{2-x}O_4$ composition is found almost in agreement with the literature values [25–28]. The decreasing trend in lattice constant (*a*), cell volume (*V*), hopping lengths (L_A and L_B), bond lengths (*A-O*, and *B-O*), and strain were observed with the Mn content (x) for them. The increasing trend in the X-ray density with increasing Mn content (x) was observed in all three systems, which was found to maintain an agreed inverse relationship with the cell volume as illustrated in **Table 1** [17, 18, 22]. The crystallite size (D_x) for the $Co_{1-x}Mn_xFe_2O_4$ composition and $CoMn_xFe_{2-x}O_4$ composition was reported to increase with increasing Mn content (x) but to decrease for $Co_{1+x}Mn_xFe_{2-x}O_4$ composition with much-enhanced value as compared to other two stoichiometric systems as illustrated in **Table 1** [17, 18, 22]. This enhancement in crystallite size and the gradual decreasing trend was due to the concurrent contribution of Co^{2+} ions in this composition and the relatively smaller ionic radius of Mn^{2+} ions and was found to follow Vegard's law as explained in the literature. The decreasing trend in particle size (PS) was noticed in $Co_{1-x}Mn_xFe_2O_4$ and $Co_{1+x}Mn_xFe_{2-x}O_4$ compositions, but its increasing trend for $CoMn_xFe_{2-x}O_4$ composition is evident in **Table 4** [17, 18, 22]. Among these three systems, the crystallite size in $Co_{1+x}Mn_xFe_{2-x}O_4$ composition is novel and significant due to the formation of granules by fudging particles together by the enhanced calcination temperature and concurrent addition of Co^{2+} with its relatively larger ionic radius (78 Å). This fact is led to changes in other physical properties of this composition and accordingly discussed in the subsequent sections to explore and/or exploit them in their possible applications.

1.2.3 Electrical properties

Cobalt ferrites are considered to be composed of layers similar to any ferrite materials. They are grain and grain boundaries [29, 30]. The electrical properties of these materials are influenced by the dopants or impurities due to change in their structural and morphological properties. These electrical properties encompass the dielectric constant or permittivity and conductivity and/or resistivity. Both the conductivity and dielectric constant have a common origin according to Koop's phenomenological theory in the manganese-doped cobalt ferrites similar to any other doped ferrite materials [17, 18, 22, 31]. The dielectric constant comes out from the polarization of the material. There are four types of polarization in the material depending on

Parameters	$Co_{1-x}Mn_xFe_2O_4$				$CoMn_xFe_{2-x}O_4$				$Co_{1+x}Mn_xFe_{2-x}O_4$			
	0.125	**0.25**	**0.375**	**0.5**	**0.125**	**0.25**	**0.375**	**0.5**	**0.125**	**0.25**	**0.375**	**0.5**
a (Å)	8.383	8.382	8.381	8.380	8.415	8.397	8.370	8.358	8.410	8.406	8.403	8.400
D (nm)	64.02	66.90	67.10	67.17	53.424	53.436	53.454	53.462	253.22	217.72	117.14	116.13
V $(Å)^3$	589.24	588.96	588.76	588.47	596.07	592.17	586.38	583.99	594.82	593.94	593.41	592.77
ρ_x (gm/cm^3)	5.104	5.107	5.109	5.111	4.750	4.779	4.824	4.842	3.351	3.355	3.358	3.362
L_A (Å)	3.630	3.629	3.627	3.623	3.644	3.636	3.624	3.619	3.64153	3.63971	3.63863	3.63733
L_B (Å)	2.964	2.963	2.962	2.960	2.975	2.969	2.959	2.955	2.97335	2.97187	2.97098	2.96992
A-O (Å)	1.817	1.815	1.813	1.811	1.822	1.818	1.812	1.809	1.82085	1.81994	1.81940	1.81875
B-O (Å)	2.097	2.095	2.092	2.089	2.103	2.099	2.092	2.089	2.10250	2.10145	2.10082	2.10007
Strain	0.00181	0.00170	0.00169	0.00168	0.00214	0.00213	0.00212	0.00211	0.14561	0.13036	0.11828	0.10833
PS (nm)	32	30	28	24	22	24	27	32	34	32	29	28

Table 4.
Structural parameters and average particle size of all three compositions [17, 18, 22].

its mechanism, namely electronic polarization, atomic (orientation) polarization, ionic polarization, and space-charge (interfacial) polarization. The conductivity or resistivity arises from the mechanism of transportation of charge carriers in the material. Cobalt ferrites are the magnetic semiconductor in which Co^{2+} ions and Fe^{3+} ions are the p-type and n-type charge carriers. In the doped cobalt ferrites, interfacial polarization plays the dominant role in their dielectric behavior and electrical conductivity due to created heterogeneity by the incorporation of dopants. The frequency and temperature response of both dielectric constant and resistivity or conductivity are instrumental to explore or exploit the possible applications in diversified fields. The frequency response of the dielectric constant exhibits the ferrimagnetic nature of the materials in all three compositions as mentioned above [17, 18, 22, 31, 32]. The increasing trend in dielectric constant with the Mn content is observed for the $Co_{1-x}Mn_xFe_2O_4$ at RT due to the decreased density of Co^{2+} ions for being replaced by Mn^{2+} ions in the B site [31]. But its unpredictable and irregular variations with Mn content for the compositions $CoMn_xFe_{2-x}O_4$ and $Co_{1+x}Mn_xFe_{2-x}O_4$ are caused by the inhomogeneity of charge careers (p-type and n-type) across the grain boundaries due to dispersed particle size distribution as reported in the literature [22–24, 33]. The electrical modulus formalism was used to all the three compositions for the purpose as (i) to identify and understand the bulk properties, electrical conductivity, and relaxation time (ii) to differentiate the grain and grain boundary conduction process from the electrode polarization effect, and (iii) microstructural correlation in both the electrical and magnetic properties of the materials. The peaks were observed in the spectra of both the real and imaginary parts of the electrical modulus for the samples of $Co_{1+x}Mn_xFe_{2-x}O_4$ composition. The frequencies corresponding to the peak in the dispersion of the real part of electrical modulus (M') divide the spectra into two wings (wing-I and wing-II) corresponding to the long-range mobility and short-range mobility of charge carriers. The critical relaxation time constant as determined from the peaks of the dispersion of the real part of the electrical modulus sets the boundary below of which the n-type (Fe^{2+}/Fe^{3+}) and above it p-type (Co^{2+}/Co^{3+}) carriers play the dominant role in their conductivity [18]. The frequency corresponding to the observed peak in the absorption of the imaginary part of electrical modulus (M″) is termed as the characteristic frequency, f_{max}, that varies with the Mn content (x). The maximum magnitude of M″ at the characteristic frequency implies the increased eddy current loss as heat radiation, and it is possible to tune this loss at any desired value by changing the Mn content. As such, it is expected that this material may be suitable to be used in hyperthermia and medical-related research. The characteristic dielectric relaxation time constant determined from the characteristic frequency depends both on the concentration level and temperature. The observed critical temperature exists in the low-temperature regime and signifies the ferromagnetic-to-spin glass state transition [18, 22]. The normal behavior of both the real part and imaginary part of complex permittivity is observed with the increase in applied frequency for the samples of $Co_{1-x}Mn_xFe_2O_4$ composition. The increasing trend in the real part of its complex permittivity in the lower frequencies is due to the decreased density of Co^{2+} ions in the B site. The minimum relaxation time constant as determined from the peaks in the spectra of the imaginary part of its complex electrical modulus shows the higher conductivity at a specific Mn content (x = 0.25) due to a higher hopping rate between F^{2+} to Fe^{3+} ions across the grain boundaries. The Nyquist plot of its complex electrical modulus identifies the dominance of carrier contributions in the conduction mechanism [18, 22]. An unpredictable variation in the value of a relative dielectric constant at Mn content (x) = 0.125 and 0.5 in the samples of $CoMn_xFe_{2-x}O_4$

composition originated from the inhomogeneity of charge careers (p-type and n-type) across the grain boundaries due to dispersed particle size distribution [17, 31]. Two semicircles appeared in the Nyquist plot of electrical modulus (M′ vs. M″) as designated by semicircle-I and semicircle-II for the samples of $CoMn_xFe_{2-x}O_4$ composition. The semicircle-I occurs in the low-frequency region and corresponds to grain-boundary contribution whereas the semicircle-II in the high-frequency region corresponding to the grain contribution [18]. The increasing trend in the estimated activation energy of the samples of $Co_{1+x}Mn_xFe_{2-x}O_4$ composition with Mn content (x) at room temperature signifies the phase tuning effect from the ferrimagnetic-to-paramagnetic phase. The temperature response of AC resistivity in the low-temperature regime for the representative sample of $Co_{1+x}Mn_xFe_{2-x}O_4$ composition showed the metallic or insulating behavior of the material. The increasing trend of activation energy with the manganese content signifies the ferrimagnetic-to-paramagnetic phase transition and thus tuning the samples of $CoMn_xFe_{2-x}O_4$ to behave as soft magnetic material [18]. The linear decreasing trend in its DC resistivity with the increase in temperature demonstrated the semiconducting behavior of the material above room temperature [17, 31]. The decreasing trend in AC resistivity of the samples of $CoMn_xFe_{2-x}O_4$ with the increase of temperature implies their semiconducting behavior of the material [18, 22]. The negative value of magnetoresistance implies the dominance of n-type charge carriers (Fe^{2+}/ Fe^{3+}) of both samples of $Co_{1-x}Mn_xFe_2O_4$ and $Co_{1+x}Mn_xFe_{2-x}O_4$ compositions in their hopping process. The complex impedance marks a single metallic band in the stoichiometric composition $Co_{1-x}Mn_xFe_2O_4$ and a double metallic band in the non-stoichiometric composition $Co_{1+x}Mn_xFe_{2-x}O_4$. This double metallic band may make this sample suitable to be used in the switching as well as actuator devices. Conversely, the higher conductivity in the single metallic band is expected to generate heat by the eddy current loss for the stoichiometric composition $Co_{1-x}Mn_xFe_2O_4$, which may also make it suitable to be used in hyperthermia and medical science-related research [23–26]. The measurement of conductance of both stoichiometric and non-stoichiometric compositions of Mn-doped cobalt ferrite nanoparticles at RT shows that "the magnitude of AC conductivity depends almost linearly with the Mn content and thus provides tunability by Mn content. This AC conductivity causes the eddy current loss while applied a varying magnetic field across the investigated sample and dissipate to heat the materials. The heat energy is expected to be selectable over the frequency band between two cutoff frequencies as determined by their 3dB points from their respective peaks and therefore make them suitable for use in temperature switching/sensing and/or thermoelectric devices. Besides the frequency exponent for both the compositions displayed that the material is a mixer of both ionic and Debye dipole-type crystals and with the increasing Mn content they tend towards more Debye dipole-type crystals" as reported in recent literature [33].

1.2.4 Magnetic properties

The AC permeability and DC magnetization are considered to be the key parameters for understanding and explaining the magnetic behavior of the material. The magnetic modulus is similarly important to separate the local behavior of defects in the material from the effects of external agents like an air gap, stray effect, etc. It helps to understand the dynamic mechanism of permeability under the influence of the AC magnetic field. A pragmatic enhancement in the real part of permeability with the concentration levels, Mn(x) over the whole frequency band is noticed for the samples of $Co_{1+x}Mn_xFe_{2-x}O_4$ composition [24]. Above room temperature, the increasing trend in the real part of permeability is caused by the dipolar orientations due to increased crystallite and/or grain size. The appearance of their relaxation peaks at around 537.5 K may be related to the spin resonance, and its fall from the peaks marks the ferromagnetic-to-paramagnetic phase transition and corresponds to the Curie temperature. Whereas the declining trend of the real part of permeability from the broadened peaks with the drop in temperature is occurred by the ceasing of dipolar orientations in the grains due to freezing effects. A crossover in the dispersion of the real part of magnetic modulus M'_m is noticed at a particular frequency around 58 kHz. Below this frequency M'_m increases but above it decreases with Mn content. The well-resolved peaks in the dispersion of the imaginary part of the magnetic modulus correspond to the resonance frequency and are found to decrease with the Mn content due to the observed decreasing trend in the crystallite size. The saturation magnetization and initial permeability are found to increase with the concentration levels x due to the antiferromagnetic effect of Mn^{2+} in the tetrahedral site according to Neel's two sublattices models [24, 33, 34]. The anomalous variation in remnant magnetization and coercivity is observed due to the migration of Co^{2+} ions from the octahedral site to the tetrahedral site during phase formation in the solid-state reaction. The remnant ratio is less than unity ($<<1$), which marks the superparamagnetic behavior of the material. The grain boundaries interact and reduce the domain wall motion that results in lower permeability at higher sintering temperatures. Below room temperature (300 K), the magnetization increases with the drop in temperature, which occurs from the dominance of effective uniaxial anisotropy constant (K_{ef}) due to the additional presence of Co^{2+} ions in the octahedral (B) site because of the non-stoichiometric composition. The coercivity decreases with a further drop in temperature from a certain peak value is the signature of the possible superparamagnetic behavior of the material. The Weiss constant depends inversely with Mn content, which signifies the ferromagnetic nature of the material in the low-temperature regime [24]. A diamagnetic behavior of the materials is also marked over the frequency band 1 kHz–500 kHz for the samples of this composition due to negative values of the real part of permeability, which is the indication of the metamaterial according to Veslago through which electromagnetic wave cannot propagate and thus may make this material suitable to be used in the magnetic shielding operation. Besides, μ-negative (MNG) and double-negative (DNG) medias are identified over the frequency band 3 MHz–120 MHz at Mn(x) = 0.5 [1]. In the low-temperature regime, the dispersion of the AC permeability shows the transformation of the ferromagnetic phase to the spin-glass state at a certain temperature peak due to the frozen tendency of spins. This peak temperature may be termed as transition temperature, which is found to vary irregularly with the Mn content in the sample of composition $CoMn_xFe_{2-x}O_4$. The remnant ratio is found to be very less than unity, which marks its superparamagnetic behavior, and may this material be suitable to be used in the spintronics applications. The frequency response of both the real and imaginary parts of AC permeability demonstrates the normal behavior of the samples of $Co_{1-x}Mn_xFe_2O_4$ compositions. A nonlinear increase in the real part of magnetic modulus (M'_m) for the samples of this composition is observed with the increase of the applied frequency that signifies the contributions of both the wall motion (wall relaxation) and the spin rotations (rotational resonance) in its magnetization [18]. Afterward, M'_m is found to a slight linear increase to a single asymptotic value, which implies the ceasing of wall motions but the only presence of spin rotations. Both the increasing and decreasing trend up to and from well-resolved peaks is observed with the increase of frequency in the dispersion or absorption of M_m''. The frequency that corresponds to the peak is known as the resonance frequency f_{res} and follows the increasing trend in the crystallite or grain size. Besides, the magnitude of M_m'' at the corresponding resonance frequency is also marked to increase with the increase in Mn content. This fact implies the more absorption of magnetic energy from the magnetic field that in turn leading to decrease spin rotations. This behavior of the material is expected to be suitable for electromagnetic suppression operations. The improvement of saturation magnetization is observed for the sample of Mn content (x = 0.5) as compared to that of un-doped cobalt ferrite nanoparticles (x = 0).

1.2.5 Comparison of electrical and magnetic properties

The applications of the synthesized materials in the real world mostly depend on the electric and magnetic properties and their tuning or alteration by the external agents like DC and/or AC electric and/or magnetic fields, electromagnetic field, the influence of temperatures both below and above room temperatures, environment, etc., thereon. A comparison is made based on the above discussion and presented in tabular form to explore the novelty of electrical and magnetic properties of manganese-doped cobalt ferrite nanoparticles of three different systems as synthesized with the composition formulas as $Co_{1+x}Mn_xFe_{2-x}O_4$, $Co_{1-x}Mn_xFe_2O_4$, and $CoMn_xFe_{2-x}O_4$ in **Table 5**:

Properties	$Co_{1+x}Mn_xFe_{2-x}O_4$	$Co_{1-x}Mn_xFe_2O_4$	$CoMn_xFe_{2-x}O_4$
Dielectric constant	The decreasing trend in the real part of the dielectric constant exhibits the ferrimagnetic nature of the material. The peaks in the absorption of the imaginary part marked the predominant grain boundary contribution to the resistivity of the material [4, 17, 22]	The frequency response of both the real and imaginary parts of the dielectric constant exhibits normal behavior. The decreasing trend in the real part of the dielectric constant exhibits the ferrimagnetic nature of the material [22]	Both dielectric constant and D-factor show the normal behavior up to 2 kHz and above 2 kHz anomalous behavior. Grain boundaries are more responsive below 10 kHz, and conductive grains become more responsive above 10 kHz that ultimately increasing the conductivity of the material almost exponentially [18, 22, 28]
Electrical modulus	A boundary is marked between the long-range and short-range mobility of charge carriers. The critical relaxation time constant set a boundary below of which n-type and above it p-type charge carriers play a predominant role in their conductivity. The lowest value of activation energy at Mn(x) = 0.375 signifies faster hopping between Fe^{3+} and Fe^{2+} ions in the	An increasing trend in the real part of electrical modulus' with the applied frequency is observed due to short-range mobility of charge carriers (p-type and n-type) because of the shorted hopping length (L_B) in the B site. No boundary is observed between the short-range and long-range mobility of charge carriers. In the low-frequency regime, the higher values of	Two semicircles (semicircle-I and II) of non-ideal shape are found to occur in the Nyquist plot of electrical modulus that separates the grain boundary and grain contributions to the electrical conductivity of the material. Semicircle-I corresponds to the grain boundary, and semicircle-II corresponds to the grain contribution to the

Properties	$Co_{1+x}Mn_xFe_{2-x}O_4$	$Co_{1-x}Mn_xFe_2O_4$	$CoMn_xFe_{2-x}O_4$
	conductive grains and is associated with the deficiencies of oxygen ions in the B-site due to additional density of Co^{2+} ions because of their non-stoichiometry [18, 22]	the imaginary part of the electrical modulus indicate lower conductivity due to interfacial polarization in the grain boundaries. The Nyquist plots of complex electrical modulus separate the contribution of n-type and p-type ions in the hoping mechanism [22]	conductivity of the material [22, 23]
Resistivity/ conductivity	The temperature response of AC resistivity in the low-temperature regime shows the metallic or insulating behavior of the material. The linear decreasing trend in the DC resistivity with the increase in temperature exhibits the semiconducting behavior of the material in the high-temperature regime. The higher rate of change of resistivity is marked with the applied magnetic field due to the combined effects of the faster hopping and higher mobility of n-type charge carriers across the grain boundaries [17, 22, 32]	The normal behavior of AC resistivity with the applied frequency is observed. The AC resistivity is found to decrease with Mn(x) below 0.01 MHz and above it independent of Mn(x). The value AC resistivity is large as compared to $CoMn_xFe_{2-x}O_4$ due to the slower mobility of Co^{2+} (p-type carriers) ions in the B site [22, 31]	The increasing trend of activation energy with the manganese content, as determined from the temperature-dependent AC conductivity curves, signifies ferrimagnetic-to-paramagnetic phase transition and thus tuning the sample to behave as soft magnetic material [18, 22, 33]
Conductance	Frequency non-responsive conductive band (FNRC) observed over the band 100 Hz–3 kHz, and frequency responsive conductive band (FRC) observed over the band 3 kHz–1 MHz [22, 31, 33]	Frequency non-responsive conductive band (FNRC) observed over the band 100 Hz–1 kHz, and frequency responsive conductive band (FRC) observed over the band 1 kHz–1 MHz [22, 32]	—
Impedance spectroscopy	The double metallic band is observed [20, 22, 26]	A single metallic band is observed [22, 29, 30]	—
Permeability	A pragmatic enhancement in the real part of permeability with Mn(x) over the 100 Hz–1 MHz. μ-negative and double negative media identified over frequency band 3MH–120 MHz at Mn(x) = 0.5. A diamagnetic behavior is marked over the frequency band 1 kHz–500 kHz at Mn (x) = 0.125 [16, 22]	The frequency response of both the real and imaginary parts of AC permeability demonstrate normal behavior. The real part of permeability is non-responsive over frequency band 550 Hz–1 MHz. The decreasing trend in relaxation time with the Mn (x) implies the damping of spins in the high-frequency range [22]	The real part of AC permeability is found to increase up to 265 Hz (snapshot) and a slight decrease over a wide band of frequencies up to around 30 MHz. Afterward, its sudden fall demonstrates anomalous behavior. In the low-temperature regime, dispersion of AC permeability shows the transformation of the ferromagnetic phase to the spin-glass state at a certain

Properties	$Co_{1+x}Mn_xFe_{2-x}O_4$	$Co_{1-x}Mn_xFe_2O_4$	$CoMn_xFe_{2-x}O_4$
			peak temperature due to the frozen tendency of spins [18, 22]
Magnetic modulus	A cross-over frequency was observed at 58 kHz. Below this frequency real part of the magnetic modulus is inversely, and above it is directly proportional to Mn (x). In the low-temperature regime, a transition of ferromagnetic phase to spin glass state is marked at 131.59 K. The critical relaxation constant increases with Mn(x) that decreases, in turn, the damping of domain walls and results in both dipolar orientations and more spin rotations to increase the permeability of the material due to change in the metal chemistry. This implies that the porosity decreased due to the additional contribution of Co^{2+} ions, which in turn removes pinning sites of the wall motion in the material [19, 22]	The resonance frequency as observed in the imaginary part of the magnetic modulus follows the increasing trend in the crystallite size. Besides, the magnitude of the imaginary part at the corresponding resonance frequency is also marked to increase with the increase in Mn (x). This fact implies the more absorption of magnetic energy from the magnetic field that in turn leading to decrease spin rotations. The relaxation time decreases with Mn (x) that signifies the increase in the damping mechanism of wall motion that causes a decrease in permeability of the material [17, 22]	The increasing trend of magnetic modulus up to around 248 kHz may be due to the collective effects of wall motion and spin rotation. Afterward, almost constancy is observed over the frequency band 248 kHz–10 MHz, which can be attributed to the spin-only rotation. The increasing trend in relaxation time with Mn(x) implies the predominance of dipolar orientations to contribute more to the permeability [18, 22]
Magnetization	Normal behavior is marked in the variation of saturation magnetization with Mn (x). In the low-temperature regime, a large saturation magnetization is observed that ranges from 120 to 240 emu g^{-1} [22]	The improvement of saturation magnetization is observed over un-doped cobalt ferrite nanoparticles [22]	Both the saturation magnetization and initial permeability decrease with Mn(x) at RT. It exhibits normal magnetization behavior [18, 22]
Coercivity	The anomalous nature of coercivity exhibits mono-domain formation with the drop in temperature at Mn (x) 0.125 and 0.375 [22]	The anomalous nature in the variation of coercivity with Mn(x) is also marked [22, 27]	The increasing trend in coercivity shows the normal magnetic behavior of this material [17, 22]
Remnant ratio	The remnant ratio was found to be below unity ($< <1$) and exhibiting possible superparamagnetic nature of the material [22]	Possible superparamagnetic behavior is observed due to the value of remnant ratio $< <1$ [17, 22, 27]	The remnant ratio is found to be very less than unity, which marks its superparamagnetic behavior [17, 22]
Curie temperature	The Curie temperature is found to be tunable with Mn (x) over the range of 535 K–565 K and is found to agree with literature values [22]		The Curie temperature is marked to be controlled by the Mn(x) content, and thus, T_C regulates magneto-mechanical hysteresis [22]

Properties	$Co_{1+x}Mn_xFe_{2-x}O_4$	$Co_{1-x}Mn_xFe_2O_4$	$CoMn_xFe_{2-x}O_4$
Magneto resistance	The negative values of magnetoresistance showed the dominance of n-type charge carriers in its conduction mechanism [10, 22, 34]	The negative values of magnetoresistance showed the dominance of n-type charge carriers in their conduction mechanism [22]	—

Table 5.
Comparison of electric and magnetic properties of three systems of Mn-doped cobalt ferrite nanoparticles.

1.2.6 The novelty

From the comparison of manganese-doped cobalt ferrite nanoparticles in three systems as seen in **Table 5**, the followings are the significances or novelty in their properties:

i. All three compositions exhibit the ferrimagnetic nature of materials and almost in agreement with the cobalt ferrites. Only the lattice parameters and crystallite size are being influenced and modified by the antiferromagnetic effect of Mn^{2+} ions, and also, the cation distribution is deviated a bit from their idealistic situation according to their initial stoichiometry.

ii. All the compositions show semiconducting behavior. But, the non-stoichiometric composition $Co_{1+x}Mn_xFe_{2-x}O_4$ exhibits either insulating or metallic behavior and semiconducting behavior, which is its novel or unique property as compared to the other two compositions.

iii. The electrical modulus itself manifests the novelty of all three compositions of manganese-doped cobalt ferrite nanoparticles wherein short-range and long-range mobility of charge carriers are distinctly identified for the composition $Co_{1+x}Mn_xFe_{2-x}O_4$. The dielectric relaxation is of single relaxation non-Debye type for $Co_{1+x}Mn_xFe_{2-x}O_4$ and $Co_{1-x}Mn_xFe_2O_4$ compositions, whereas it is a double relaxation type for $CoMn_xFe_{2-x}O_4$ composition.

iv. Impedance spectroscopy shows the double metallic band for the $Co_{1+x}Mn_xFe_{2-x}O_4$ composition, whereas a single metallic band is observed in $Co_{1-x}Mn_xFe_2O_4$. These are unique for these two systems, and therefore, it may be regarded as the novel property.

v. A diamagnetic behavior manifests μ-negative (MNG) and double-negative (DNG) media in $Co_{1+x}Mn_xFe_{2-x}O_4$, which are absent in the other two compositions and therefore unique or novel property.

vi. The anomalous magnetic behavior in respect of coercivity and remnant magnetization is unique for the samples of $Co_{1+x}Mn_xFe_{2-x}O_4$ compositions below the room temperature as compared to the other two compositions.

vii. The transformation of the ferromagnetic phase to the spin-glass state is another important novel property of $Co_{1+x}Mn_xFe_{2-x}O_4$ and $CoMn_xFe_{2-x}O_4$ compositions.

viii. The magnetic modulus manifests that the dependence of permeability on manganese constant that reverses at a particular frequency termed as crossover frequency, which is unique for $Co_{1+x}Mn_xFe_{2-x}O_4$ composition.

ix. The decreasing trend in the resonance frequency as determined from the well-resolved peaks of the imaginary part of the magnetic modulus with increasing Mn(x) for $Co_{1+x}Mn_xFe_{2-x}O_4$ follows the decreasing trend of its crystallite size that manifest the correlation between permeability and crystallite or grain size and therefore novel behavior of the material.

x. The increasing trend in resonance frequency with Mn(x) for the $Co_{1-x}Mn_xFe_2O_4$ follows the increasing trend of its crystallite size and shows the novel behavior.

xi. The tunability of the Curie temperature by manganese content in $CoMn_xFe_{2-x}O_4$ facilitates control over the mganetomechanical hysteresis of the material.

1.3 Possible applications

The novel behavior of manganese-doped cobalt ferrite nanoparticles in three compositions as discussed above may make them suitable to be used in the following possible applications:

i. The semiconducting nature of the material may make them suitable to be used in thermoelectric devices.

ii. The decreasing trend in the coercivity of $Co_{1+x}Mn_xFe_{2-x}O_4$ composition may make this material suitable to be used as a ferrite core in the high-frequency inductor, transformer, and various microelectronic devices

iii. The appearance of peaks in the dispersion of the imaginary part of the dielectric constant shows the grain boundary contribution to the resistivity in $Co_{1+x}MnxFe_{2-x}O_4$ which in turn corresponds to the thermal energy dissipation. This behavior of dielectric constant may make this material suitable for use in hyperthermia or related research in medical science.

iv. The equality in the real part and imaginary parts of complex permittivity at resonance suggests this material of $Co_{1+x}MnxFe_{2-x}O_4$ composition may be used in high-frequency switching devices.

v. The enhanced negative magnetoresistance and its variation nature make the materials of $Co_{1+x}MnxFe_{2-x}O_4$ composition be used to identify and control the electron transport properties in the sensor and spintronics applications.

vi. The varied nature of temperature-dependent resistivity may make the material of $Co_{1+x}MnxFe_{2-x}O_4$ composition suitable to be used in the high-temperature sensor devices.

vii. The dispersion of the real part of permeability is found almost independent over the frequency band 550 Hz–1 MHz, and the imaginary part shows non-responsive over frequency band 1.5 kHz–1 MHz and makes this material of $Co_{1-x}Mn_xFe_2O_4$ composition a good candidate to be used in high-frequency applications.

viii. The increasing nature of magnetization with the temperature at constant concentration levels x provides the ability to control and probe the magnetization of the material and thus make the system suitable to be used in the field of sensor application.

ix. The metallic frequency band as observed at x = 0.5 of all three systems in their anomalous zone above 1 MHz may make the material suitable to be used in the radio frequency detectors and fast switching devices.

x. This negative values of μ' are the indication of a possible metamaterial that is capable of distorting or absorbing the electromagnetic waves, which makes this material suitable to be used in electromagnetic suppression or shielding operation. Conversely, the constancy in μ' and ?′ over a frequency band 500 kHz–100 MHz is an indication of potential technological material for patch/dielectric antenna.

xi. The increasing nature of magnetization with the temperature at constant concentration levels x provides the ability to control and probe the magnetization of the material and thus make the system suitable to be used in the field of sensor applications.

xii. The decreasing rate of resistivity with the applied magnetic field at room temperature may make this sample suitable to be used in the field of position and biosensors.

xiii. The relative enhancement in saturation magnetization and initial permeability in $Co_{1-x}Mn_xFe_2O_4$ and $CoMn_xFe_{2-x}O_4$ as compared to that in un-doped cobalt ferrite nanoparticles reveals the use of Mn in fine-tuning the new magnetic materials for technological applications in microelectronic devices.

xiv. The tunability of the Curie temperature of the material of composition $CoMn_xFe_{2-x}O_4$ by the Mn(x) content may make this sample suitable to be used in the magneto-mechanical sensors.

2. Conclusion

From the above discussion, it is interesting to note that, creating defects intentionally in a methodic way, the physical properties will be changed, which is the crux route to tune or tailor the properties in the solids. Intensive investigations, therefore, were carried out over the years, and their outcomes have put in practical applications in diversified fields. Manganese-doped cobalt ferrite irrespective of its bulk and nano-dimension forms has already been utilized in applications starting from the electronic industry to biomedical research, diagnostic and labeling, and targeted delivery of medicines. In this chapter, some novel approach like electrical modulus, magnetic modulus, and impedance spectroscopy has been used to identify the change and new properties of manganese-doped cobalt ferrite nanoparticles and highlighted some possible applications that include hyperthermia and medical-related research, magnetic field shielding/suppression operation, etc. The tunability of AC conductivity of both stoichiometric and non-stoichiometric compositions of Mn-doped cobalt ferrite nanoparticles is suggested to be suitable for use in thermoelectric devices. However, there is still room for further investigations more rigorously, a few of them are suggested here for the researchers, scientists, engineers, and technologists to open up their thoughts across the globe:

i. The photocatalytic conversion of visible solar energy to generate e^{-}/h^{+} in turn may produce reactive oxygen species through the redox process for the degradation of pollutants in water.

ii. In a quest of exploring new alternatives of RE permanent magnets by investigating non-stoichiometric manganese-doped cobalt ferrites and compositing them to nickel to enhance magnetism and coercivity.

iii. An investigation of this material is required to enhance the dye degradation capability.

iv. Improvement of the efficiency and savings of energy in renewable energy devices using cobalt ferrite nanoparticles with non-stoichiometric composition.

Acknowledgements

The authors are thankful to the International Science Programs (ISP), Uppsala University, Sweden, for financial and technical support, also to the Department of Physics, Bangladesh University of Engineering and Technology, and the Department of Physics, Military Institute of Science and Technology for experimental support.

Abbreviations

Co	symbol for cobalt
Mn	symbol for manganese
Fe	SYMBOL of iron
RT	Room temperature
x	signifies Mn concentration/concentration levels at weight percentage
Å	Angstrom (unit of dimension) = 10^{-10} m
K	Kelvin scale of temperature
AC	alternating current
nm	nanometer (unit of dimension) = 10^{-9} m
emu g^{-1}	electromagnetic unit per gram (unit of magnetization

XRD	X-ray diffraction
FESEM	field emission electron microscopy
M (H)	magnetization as a function of magnetic field
M_S	saturation magnetization
M_R	remnant magnetization
H_C	coercivity
μ_i	initial permeability
n_B	magnetic moment per formula unit in Bohr Magneton
μ'	real part of AC permeability
μ''	imaginary part of permeability
χ	susceptibility
$\tan\delta$	loss factor (corresponds to both dielectric and magnetic loss)
ε'	real part of permittivity
ε''	imaginary part of permittivity
ε_r	relative dielectric constant
C	capacitance
Z	impedance
X	reactance
R	resistance
ρ_{ac}	AC resistivity
σ_{ac}	AC conductivity
f_o	resonance frequency
τ	relaxation time constant
f_c	critical frequency
f_{max}	maximum characteristic frequency
M'	real part of electric modulus
M''	imaginary part of electric modulus
M_m'	real part of magnetic modulus
M_m''	imaginary part of magnetic modulus
$\Delta M(T)$	change of magnetization as a function of temperature
ΔT	change in temperature
ΔH	change in magnetic field
ΔS	change in entropy
ΔS_m	change in magnetic entropy
ΔS_{lat}	change in entropy due to lattice vibrations
ΔS_{elec}	change in entropy due to electronic contribution

Appendix: used formulas

The formulas used for calculation of various parameters for analysis of structural, magnetic, and electrical properties along with software are listed in **Table A.1**.

Parameters	Formula	Remarks
Structural		
Lattice constant	$d = \frac{a}{\sqrt{h^2+k^2+l^2}}$	Using High Score Plus software
Crystallite size	$0.9\lambda/\beta\cos\theta$	Debye–Scherrer formula

Parameters	Formula	Remarks
Tetrahedral hoping length	$(L) = \frac{a\sqrt{3}}{4}$A	Stanley's equation
Octahedral hoping length	$[L] = \frac{a\sqrt{2}}{4}$A	
Tetrahedral bond length	$A - O = \left(u - \frac{1}{4}\right)a\sqrt{3}$A	
Octahedral bond length	$B - O = \left(\frac{5}{8} - u\right)\mathring{a}$A	
X-ray density	$\rho_x = ZM/Na^3$	
Porosity	$P\% = \left(1 - \frac{\rho_x}{\rho}\right) \times 100$	
Strain	$\beta \cos\theta = \frac{k\lambda}{D} + 4\epsilon \sin\theta$	Williamson–Hall equation (plot)
Magnetic		
Magnetization	$(V_{measd} \times Cac - Const)/M_{measrd}$	Using VSM
Coercivity	$(H_{C1} + H_{c2})/2$	Using hysteresis curves
Remnant magnetization	$(M_{C1} + M_{c2})/2$	
Curie temperature	Extrapolated intercept value	Using M–T curves
Anisotropy constant		
Initial permeability	The slope value at initial magnetization	Using M–H curves
Susceptibility	$M/\mu_0 H$	
Electrical		
DC resistivity	$\rho_{dc} = RA/l$	Using the 4-probe method
DC conductivity	$\sigma_{dc} = 1/\rho_{dc}$	
AC resistivity	$\rho_{dc} = RA/l$	Using Wynekerr impedance analyzer
AC conductivity	$\sigma_{dc} = 1/\rho_{dc}$	
Magnetoresistance	$MR\% = \frac{\rho_H - \rho_0}{\rho_0} \times 100$	Using the 4-probe method
Electromagnetic		
Complex permittivity	$\varepsilon = \varepsilon' - j\varepsilon''$	Measurement of real and imaginary parts by impedance analyzer
Complex permeability	$\mu = \mu' - j\mu''$	
Impedance	$Z = \sqrt{R^2 + X^2}$	
Characteristics impedance	$Z_0 = \sqrt{\mu'/\varepsilon'}$	
Miniaturization factor	$n = \sqrt{\mu' \times \varepsilon'}$	
Dielectric constant	$\varepsilon_r = \frac{C}{C_0}$	Measurement of capacitance by impedance analyzer
Dielectric loss	$\tan\delta = \varepsilon''/\varepsilon'$	Measurement of real and imaginary parts by impedance analyzer
Magnetic loss	$\tan\delta = \mu''/\mu'$	

Parameters	Formula	Remarks
Eddy current loss	$W_{ed.loss} = \omega\mu_r\mu_0 H^2 V \sin\delta$	Usual meanings
Currie constant, C	The slope of $\frac{1}{\chi} - T$ plot	Usual meanings
Weise constant, θ	The intercept at the temperature axis	
Electric modulus	$M'(\omega) = \frac{\varepsilon'(\omega)}{\varepsilon'(\omega)^2 + \varepsilon''(\omega)^2}$ $M''(\omega) = \frac{\varepsilon''(\omega)}{\varepsilon'(\omega)^2 + \varepsilon''(\omega)^2}$	Real part Imaginary part
Magnetic modulus	$M'_m(\omega) = \frac{\mu'(\omega)}{\mu'(\omega)^2 + \mu''(\omega)^2}$ $M''_m(\omega) = \frac{\mu''(\omega)}{\mu'(\omega)^2 + \mu''(\omega)^2}$	Real part Imaginary part

Table A.1.
The formula/procedure used for calculation of different parameters used in this thesis paper [18, 19, 22].

Author details

Md. Ziaul Ahsan
Department of Chemistry and Physics, Gono Bishwabidyalay, Dhaka, Bangladesh

*Address all correspondence to: ziaul8808828@gmail.com

IntechOpen

© 2023 The Author(s). Licensee IntechOpen. This chapter is distributed under the terms of the Creative Commons Attribution License (http://creativecommons.org/licenses/by/3.0), which permits unrestricted use, distribution, and reproduction in any medium, provided the original work is properly cited.

References

[1] Vasundhara K, Achary SN, Deshpande SK, Babu PD, Meena SS, Tyagi AK. Size-dependent magnetic and dielectric properties of nano $CoFe_2O_4$ prepared by a salt assisted gel-combustion method. Journal of Applied Physics. 2013;**113**:194101

[2] Behera C, Choudhary RNP, Piyush R. Size effect on electrical and magnetic properties of mechanically alloyed $CoFe_2O_4$ nanoferrite. Journal of Material Science. 2015;**2**:2343-2356

[3] Ahsan Z, Khan FA, Islam A. Influence of milling time on structural and electromagnetic properties of manganese doped cobalt ferrite nanoparticles. International Journal of Materials Science and Applications. 2018;**7**(6): 199-208

[4] Ahsan MZ, Khan FA. Structural and electrical properties of manganese doped cobalt ferrite nanoparticles. Materials Science Nanotechnology. 2018;**2**(2):19

[5] Jauhar S, Kaur J, Goyal A, Singhal S. Tuning the properties of cobalt Ferrite: A road towards diverse applications. RSC Advances. 2020;**2020**. DOI: 10.1039/C6RA21224G

[6] Wajea SB, Hashim M, Ismail I. Effects of sintering temperature on grain growth and the complex permeability of $Co_{0.2}Ni_{0.3}Zn_{0.5}Fe_2O_4$ material prepared using mechanically alloyed nanoparticles. Journal of Magnetic Materials. 2011;**323**:1433-1439

[7] Ahmad R, Gul HI, Zarrar M, Anwar H, Khan MB, Khan A. Improved electrical properties of cadmium substituted cobalt ferrites nano-particles for microwave application. Journal of Magnetic Materials. 2016; **405**:28

[8] Kolekar YD, Sanchez L, Rubio EJ, Ramana CV. Grain and grain boundary effects on the frequency and temperature dependent dielectric properties of cobalt ferrite-hafnium composites. Solid State Communication. 2014;**184**:34

[9] Sivakumar N, Narayanasamy A, Chinnasamy CN, Jeyadevan B. Influence of thermal annealing on the dielectric properties and electrical relaxation behaviour in nanostructured $CoFe_2O_4$ ferrite. Journal of Physics Condensened Materials. 2007. DOI: 10.1088/0953-8984/19/38/386201

[10] Ponpandian N, Balaya P, Narayanasami A. Electrical conductivity and dielectric behaviour of nanocrystalline $NiFe_2O_4$ spinel. Journal of Physics Condensened Materials. 2002; **14**:3221

[11] Singh JP, Kumar H, Singhal A, Sarin N, Srivastava RC, Chae KH. Solubility limit, magnetic interaction and conduction mechanism in rare earth doped spinel ferrite. Applied Science Letters. 2016. DOI: 10.17571/appslett.2016.02001

[12] Kumar H, Srivastava RC, Singh JP, Negi P, Agrawal HM, Das D, et al. Structural and magnetic study of dysprosium substituted cobalt ferrite nanoparticles. Journal of Magnetic Materials. 2016;**401**:16

[13] Ahsan MZ, Khan FA. Study of structural, electrical and magnetic properties of manganese doped cobalt ferrite nanoparticles with non-stoichiometric composition. Journal of Physical Science and Application. 2017; 7(6):30-37. DOI: 10.17265/2159-5348/2017.06.005

[14] Ahsan Z, Islam A. A theoretical approach: Effects of Mn substitution in cobalt ferrite. American Journal of Applied Scientific Research. 2019;**5**(3): 56-61

[15] Ahsan MZ, Khan FA, Ahsan PA, Islam MA. Exploiting of magnetocaloric effect from manganese doped cobalt ferrite nanoparticles for low-temperature applications. Invention Journal of Research Technology in Engineering & Management (IJRTEM). 2018;**2**(12):43-46

[16] Ahsan MZ, Khan FA, Islam MA, TanzinaTabassum M, Alam K. Study on AC permeability and permittivity of manganese doped cobalt ferrite nanoparticles. Journal of Physical Communication. 2018;**2018**:2105008. DOI: 10.1088/2399-6528/aade79

[17] Ahsan MZ, Khan FA, Islam MA. Frequency and temperature dependent dielectric and magnetic properties of manganese doped cobalt ferrite nanoparticles. Journal of Electrical Materials. 2019;**48**:7721-7729. DOI: 10.1007/s11664-019-07583-y

[18] Ahsan MZ, Khan FA, Islam MA. Frequency and temperature dependent intrinsic electric properties of manganese doped cobalt ferrite nanoparticles. Results in Physics. 2019; **14**:102484

[19] Ahsan MZ, Islam MA, Bally AA, Khan FA. Spectroscopic analysis for electric and magnetic properties of manganese doped cobalt nanoferrite. Results in Physics. 2020;**17**:103172

[20] Ahsan MZ, Islam MA, Bally MAA, Tabassum T, Khan FA. Significances of complex impedance in stoichiometric and non-stoichiometric compositions of manganese doped cobalt nanoferrites. International Journal of Innovative Studies in Sciences and Engineering Technology. 2020;**6**(7)

[21] Abdel Maksoud MIA, El-Sayyad GS, Elkodous MA, Awed AS. Controllable synthesis of Co1x $MxFe_2O_4$ nanoparticles (M = Zn, Cu, and Mn; x = 0.0 and 0.5) by cost-effective sol–gel approach: Analysis of structure, elastic, thermal, and magnetic properties. Journal of Materials Science: Materials in Electronics. 2020. DOI: 10.1007/s10854-020-03518-0

[22] Ahsan MZ. Structural correlation and tunability of magnetic and electric properties of manganese doped cobalt ferrite nanoparticles [thesis]. 2020

[23] Yakubu A, Abbas Z, Ibrahim NA, Hashim M. Effect of temperature on structural, magnetic and dielectric properties of cobalt ferrite nanoparticles prepared via co-precipitation method. PSIJ. 2015;**8**:18787

[24] Valenzuela R. Novel applications of ferrites. Physics Research International. 2012;**591**:9

[25] Valenzuela R. Magnetic Ceramics. Cambridge: Cambridge University Press; 1994

[26] Kharisov BI, González CMO, Kharissova OV , Gonzalez LT, Méndez YP, Uflyand IE. Chapter 2: Ferrite nanoparticles in the modern catalytic processes. In: Varma RS, Banerjee B, editors. Industrial Applications. Berlin, Boston: De Gruyter; 2022. pp. 37-72. DOI: 10.1515/9783110782165-002

[27] Adeela N, Khan M, Khan U, Iqbal M, Riaz S, Naseem S. Temperature-dependent magnetic response of antiferromagnetic doping in cobalt ferrite nanostructures. Nanomaterials. 2016;**6**:1-13

[28] Tamiya S, Ashwani K, Shrivastava S. Effect of Mn substitution on structural and magnetic properties of cobalt ferrite. Advanced Materials Proceedings. 2017; **2**(9):547-551

[29] Purnama B, Tri Wijayanta A, Suharyana B. Effect of calcination temperature on structural and magnetic properties in cobalt ferrite nanoparticles. Journal of King Saud University-Science. 2019;**31**:956-960

[30] Jian Chin C, Wen-Chng C, Yin-Chun T, Chi-Jen S. Effect of calcination temperature on the crystallite growth of cerium oxide nanopowders prepared by the co-precipitation process. Journal of Alloys and Compounds. 2010;**496**(1): 364–369

[31] Shamima N, Chowdhury F-U-Z, Hoque SM. Study of hyperthermia temperature of manganese-substituted cobalt nano ferrites prepared by chemical co-precipitation method for biomedical application. Journal of Magnetic Materials. 2019;**479**:126-134

[32] Kuanr K, Srivastava GP. Dispersion observed in electrical properties of titanium-substituted lithium ferrites. Journal of Applied Physics. 1994;**75**(10): 6116-6118

[33] Ahsan MZ, Islam MA, Khan FA. Tunability of ac conductivity in manganese doped cobalt ferrite nanoparticles. Results in Physics. 2020; **2020**. DOI: 10.1016/j.rinp.2020.103782

[34] Paulsen JA, Ring AP, Lo CCH, Synder JE, Jiles DC. Manganese-substituted cobalt ferrite magnetostrictive materials for magnetic stress sensor applications. Journal of Applied Physics. 2005;**97**:044502